Meet the editor

Carlos Frajuca graduated with a higher education degree in mechanics from Sao Carlos Federal University, Brazil, in 1986, a Ph.D. in Science from Sao Paulo University, Brazil, in 1996, and a two-year research period at Louisiana State University, USA, working with electromechanical transducers on the detection of Gravitational Waves. He obtained a habilitation degree in Stellar Astrophysics, Compact Objects and Cosmology from the Sao Paulo Federal University, Brazil. He is a member of master's programs at the Rio Grande Federal University (FURG) and at the Sao Paulo Federal Institute (IFSP), where he was the director of teaching, research, and postgraduate studies. He received research productivity scholarships from the National Council for Scientific and Technological Development (CNPq) in 2014, 2018, and 2022. He was assigned to the Brasilia Federal Institute in 2009 as Dean of Research and Deputy Rector.

Contents

Preface

This book is a useful reference for those wanting to understand gravitational waves. It is organized into two sections on the theory of gravitational waves and observations.

Section one contains five chapters. Chapter 1, "Gravitational Waves and Parametrizations of Linear Differential Operators", introduces the topic. Chapter 2, "Gravitational Waves, Fields, and Particles in the Frame of (1 + 4)D Extended Space Model", presents gravitational waves in an extended space model in five dimensions. Chapter 3, "Analytical Description of Unified Field Theory for Electromagnetic and Gravity Fields with the Introduction of Quantized Spacetime and Zero-Point Energy", describes gravitational waves in a quantized space-time with five dimensions (four space coordinates and one time coordinate). Chapter 4, "New Conservation Law as to Hubble Parameter, Squared Divided by Time Derivative of Inflaton in Early and Late Universe, Compared with Discussion of HUP in Pre Planckian to Planckian Physics, and Relevance of Fifth Force Analysis to Gravitons and GW", describes a proposed new conservation law effects in cosmology, gravitational waves, and a possible fifth force. Chapter 5, "Kantowski-Sachs Barrow Holographic Dark Energy Model in Saez-Ballester Theory of Gravitation", describes dark energy and some aspects of gravitational waves in a different theory of gravity. The Saez-Ballester theory of gravitation is considered to be the proper theory to study dark energy and the accelerated universe.

Section 2 includes two chapters. Chapter 6, "Main Experiments for Detection of Gravitational Waves at Frequency below 3 kHz: A Quick Review", summarizes experiments for detecting gravitational waves with frequencies below 3 kHz. Finally, Chapter 7, "Effects of Gravitational Waves on Two-Level Atom Moving in a Quantized Traveling Light Field: Exact Solution via Path Integral", discusses the effects of gravitational waves over atoms, which can be used for gravitational wave detection.

Carlos Frajuca
IMEF, Rio Grande Federal University,
Rio Grande, Brazil

Section 1

Theory

Chapter 1

Gravitational Waves and Parametrizations of Linear Differential Operators

Jean-Francois Pommaret

Abstract

When $\mathcal{D} : \xi \to \eta$ is a linear differential operator, a "direct problem" is to find the generating compatibility conditions (CC) in the form of an operator $\mathcal{D}_1 : \eta \to \zeta$ such that $\mathcal{D}\xi = \eta$ implies $\mathcal{D}_1\eta = 0$. Similarly, $\mathcal{D}_1\eta = \zeta$ may imply $\mathcal{D}_2\zeta = 0$ and so on. Conversely, when $\mathcal{D}_1$ is given, a much more difficult "inverse problem" is to look for an operator $\mathcal{D} : \xi \to \eta$ with generating CC $\mathcal{D}_1\eta = 0$. If this is possible, one shall say that the operator $\mathcal{D}_1$ is parametrized by $\mathcal{D}$. The parametrization is "minimum" if the differential module defined by $\mathcal{D}$ does not contain any free differential submodule. The systematic use of the adjoint of a differential operator provides a constructive test. The parametrization of the Cauchy stress operator in arbitrary dimension n has attracted many famous scientists (G.B. Airy in 1863 for $n = 2$, J.C. Maxwell in 1863, G. Morera and E. Beltrami in 1892 for $n = 3$, A. Einstein in 1915 for $n = 4$). We prove that all these works are already explicitly using the self-adjoint Einstein operator, *which cannot be parametrized*, and are thus all based on a confusion between the Cauchy operator, (adjoint of the Killing operator $\mathcal{D}$), and the *div* operator induced from the Bianchi operator $\mathcal{D}_2$ CC of the Riemann operator $\mathcal{D}_1$ parametrized by $\mathcal{D}$. This purely mathematical result deeply questions the origin and existence of gravitational waves that are solutions of the adjoint of the Ricci operator. We do believe that Einstein was aware of these previous works as the comparison needs no comment. The same methods are also used in order to revisit the mathematical foundations of electromagnetism.

Keywords: differential sequence, Killing operator, Riemann operator, Bianchi operator, general relativity, gravitational waves, Maxwell equations

1. Introduction

The problem of parametrizing the Einstein operator or, equivalently and by analogy with Maxwell equations for electromagnetism (EM), to decide about the existence of a potential for Einstein equations in vacuum, has been proposed for the first time as a 1000 dollars challenge by J. Wheeler while the author of this paper was a visiting student of D. C. Spencer in 1970 at Princeton university.

No progress at all has been done during the next 25 years, till the author gave a negative answer in 1995, contrary to what the general relativity (GR) community was believing [1, 2]. Indeed, after teaching elasticity for 25 years to high-level students in some of the best french civil engineering schools, the author of this paper proposed an exercise explaining why a dam made with concrete is *always* vertical on the water side with a slope of about 42 degrees on the other free side in order to obtain a minimum cost and the auto-stability under cracking of the surface underwater (See ([3], p. 108) and the introduction of [4] for more details). Surprisingly, the main tool involved is the approximate computation of the Airy function inside the dam in this two-dimensional elasticity problem. The author discovered at that time that no one of the other teachers did know that the Airy parametrization was nothing else than the adjoint of the linearized Riemann operator used as generating CC for the deformation tensor by any engineer. Being involved in GR with A. Lichnerowicz at that time, he got for the first time the idea of using the adjoint of an operator in a systematic way. Giving a seminar in Paris in order to present this result, somebody in the audience told him about a possible link with the recently published master thesis of the Japanese student M. Kashiwara [5]. It has been a shock to discover this mixing up of differential geometry [6, 7] and homological algebra [8–10], now called "*Differential Homological Algebra*", in particular the introduction of the *Differential Extension Modules* (See [4, 5, 11–13] for extensive references) (See also Zbl 1079.93001 for comments). It is only recently that he discovered GR could be considered as a way to parametrize the Cauchy operator and to introduce gravitational waves (GW) [14, 15]. It follows that *exactly the same confusion* has been done by Maxwell, Morera, Beltrami and Einstein because, *in all these cases*, the operator considered is self-adjoint. However, most of the pure mathematicians involved were proud not to be interested by applications and in any case unable to do any computation. Accordingly and until now, the GR community has *never* wanted to take these new tools into account and Ref. [16] is providing a good example of such a poor situation. This is the reason for which we have not been able to provide *any* other reference and why all the results presented are new.

For example, the fact that the Cauchy operator is the adjoint of the Killing operator for the Euclidean metric is *apparently* in the chapter "variational calculus" of any textbook of continuum mechanics, and the parametrization problem has been quoted by many famous authors, as we said in the Abstract, but only from a computational point of view. The same comment can be done for the two sets of Maxwell equations in electromagnetism [4, 13]. However, it is still not known that the adjoint of the 20 components of the Bianchi operator has been introduced by C. Lanczos between 1939 and 1962 [17] as we explained with details in [18] by using Spencer cohomology. The main trouble is that these two problems have *never* been treated in an intrinsic way and, in particular, changes of coordinates have *never* been considered. The same situation can be met for the Vessiot structure equations but is out of the scope of this paper [12, 19].

Lemma 1.1. When $y^k = f^k(x)$ is invertible with $\Delta(x) = det\left(\partial_i f^k(x)\right) \neq 0$ and inverse $x = g(y)$, then we have *n identities* $\frac{\partial}{\partial y^k}\left(\frac{1}{\Delta}\partial_i f^k(g(y))\right) = 0$ (See [4] p 490 and [20] for other applications).

Proof: Using the chain rule for derivatives, we get $\partial_i \Delta = \partial_{ij} f^k$ cofactor $\left(\partial_i f^k\right) = \partial_{ij} f^k \Delta \frac{\partial g^j}{\partial y^k}$ and thus:

$$\frac{1}{\Delta}\partial_{ij}f^k\frac{\partial g^j}{\partial y^k}-\frac{1}{\Delta^2}\partial_i f^k\left(\frac{\partial g^j}{\partial y^k}\partial_j\Delta\right)=\frac{1}{\Delta}\partial_{ij}f^k\frac{\partial g^j}{\partial y^k}-\frac{1}{\Delta^2}\partial_i\Delta=0$$

□

Proposition 1.2. The Cauchy operator is the adjoint of the Killing operator in arbitrary dimension n, up to sign. Similarly, when $n=4$, the Maxwell operator $\wedge^4T^*\otimes\wedge^2T\overset{ad(d)}{\rightarrow}\wedge^4T^*\otimes T:(\mathcal{F}^{ij})\rightarrow(\partial_i\mathcal{F}^{ij}=\mathcal{J}^j)$ is the adjoint of the parametrizing operator $T^*\overset{d}{\rightarrow}\wedge^2T^*:A\rightarrow dA=F$ in electromagnetism (EM), *independently of the Minkowski constitutive relations* $F\rightarrow\mathcal{F}$.

Proof: Let X be a manifold of dimension n with local coordinates $(x^1,\ldots,x^n)$, tangent bundle T, and cotangent bundle T^*. If $\omega\in S_2T^*$ is a metric with $det(\omega)\neq 0$, we may introduce the standard L, that is, derivative in order to define the first order Killing operator $\xi\rightarrow\mathcal{L}(\xi)\omega$, namely:

$$\mathcal{D}:\xi\in T\rightarrow\Omega=\left(\Omega_{ij}=\omega_{rj}(x)\partial_i\xi^r+\omega_{ir}(x)\partial_j\xi^r+\xi^r\partial_r\omega_{ij}(x)\right)\in S_2T^* \quad (1)$$

Here starts the problem because, in our opinion at least, a systematic use of the (formal) adjoint operator has *never* been done in mathematical physics (continuum mechanics, EM, ...) apart from a variational procedure. As will be seen later on, the purely intrinsic definition of the adjoint can only be done in the theory of differential modules by means of the so-called *side-changing functor*. From a purely differential geometric point of view, the idea is to associate to any vector bundle E over X, a new vector bundle $ad(E)=\wedge^nT^*\otimes E^*$, where E^* is obtained from E by patching local coordinates while inverting the transition matrices, exactly like T^* is obtained from T in tensor calculus. It follows that the stress tensor $\sigma=(\sigma^{ij})\in ad(S_2T^*)=\wedge^nT^*\otimes S_2T$ is *not* a tensor but a tensor density, that transforms like a tensor up to a certain power of the Jacobian matrix. When $n=4$, the fact that such an object is called stress-energy tensor does not change anything as it cannot be related to the Einstein tensor which is a true *tensor* indeed. Of course, it is always possible in GR to use $(det(\omega))^{\frac{1}{2}}$ but, as we shall see, the study of contact structures *must* be done without any reference to a background metric. In any case, we may define as usual:

$$ad(\mathcal{D}):\wedge^nT^*\otimes S_2T\rightarrow\wedge^nT^*\otimes T:\sigma\rightarrow\varphi \quad (2)$$

Multiplying Ω_{ij} by σ^{ij} and integrating by parts, the factor of $-2\xi^k$ is easily seen to be:

$$\nabla_i\sigma^{ik}=\partial_i\sigma^{ik}+\gamma^k_{ij}\sigma^{ij}=\varphi^k \quad (3)$$

with well known Christoffel symbols $\gamma^k_{ij}=\frac{1}{2}\omega^{kr}(\partial_i\omega_{rj}+\partial_j\omega_{ir}-\partial_r\omega_{ij})$.

However, if the stress should be a tensor, we should get for the covariant derivative:

$$\nabla_r\sigma^{ij}=\partial_r\sigma^{ij}+\gamma^i_{rs}\sigma^{sj}+\gamma^j_{rs}\sigma^{is}\Rightarrow\nabla_i\sigma^{ik}=\partial_i\sigma^{ik}+\gamma^r_{ri}\sigma^{ik}+\gamma^k_{ij}\sigma^{ij}$$

The difficulty is to prove that we do not have a contradiction because σ is a tensor density.

If we have an invertible transformation like in the lemma, we have successively by using it:

$$\tau^{kl}(f(x)) = \frac{1}{\Delta}\partial_i f^k(x)\partial_j f^l(x)\sigma^{ij}(x)$$

$$\frac{\partial\tau^{kl}}{\partial y^k} = \left(\frac{\partial}{\partial y^k}\left(\frac{1}{\Delta}\partial_i f^k\right)\right)\partial_j f^l\sigma^{ij} + \frac{1}{\Delta}\partial_i f^k\frac{\partial}{\partial y^k}\left(\partial_j f^l\right)\sigma^{ij} + \frac{1}{\Delta}\partial_i f^k\partial_j f^l\frac{\partial}{\partial y^k}\sigma^{ij}$$

$$\Rightarrow \frac{\partial\tau^{ku}}{\partial y^k} = \frac{1}{\Delta}\left(\partial_{ij}f^u\right)\sigma^{ij} + \frac{1}{\Delta}\partial_j f^u\partial_i\sigma^{ij}$$

Now, we recall the transformation law of the Christoffel symbols, namely:

$$\partial_r f^u(x)\gamma^r_{ij}(x) = \partial_{ij}f^u(x) + \partial_i f^k(x)\partial_j f^l(x)\bar{\gamma}^u_{kl}(f(x))$$

$$\Rightarrow \frac{1}{\Delta}\partial_r f^u\gamma^r_{ij}\sigma^{ij} = \frac{1}{\Delta}\partial_{ij}f^u\sigma^{ij} + \bar{\gamma}^u_{kl}(y)\tau^{kl}$$

Eliminating the second derivatives of f, we finally get:

$$\psi^u = \frac{\partial\tau^{ku}}{\partial y^k} + \bar{\gamma}^u_{kl}\tau^{kl} = \frac{1}{\Delta}\partial_r f^u\left(\partial_i\sigma^{ir} + \gamma^r_{ij}\sigma^{ij}\right) = \frac{1}{\Delta}\partial_r f^u\varphi^r$$

This tricky technical result, which is not evident, explains why the additional term we had is just disappearing, in fact, when σ is a density.

The case of EM is even simpler because $\partial_{ij}f^u\mathcal{F}^{ij} = \partial_{ji}f^u\mathcal{F}^{ji} = -\partial_{ij}f^u\mathcal{F}^{ij} = 0$ and γ is not needed. *The two sets of Maxwell equations are thus separately invariant by any diffeomorphism*. Though surprising it may look like, the conformal group of space–time is only the maximum group of invariance of the Minkowski constitutive law in vacuum. Indeed, this law is *not at all* $F_{ij} = \mu_0\omega_{ir}\omega_{js}\mathcal{F}^{rs}$, where μ_0 is the magnetic constant because such a relation is *not* tensorial as F is a 2-form, that is, a 2-covariant tensor, but $\mathcal{F}$ is a 2-contravariant tensor *density*. Hence, introducing the *metric density* $\hat{\omega}_{ij} = (|\det(\omega)|)^{-1/n}\omega_{ij}$, we must set $F_{ij} = \mu_0\hat{\omega}_{ir}\hat{\omega}_{js}\mathcal{F}^{rs}$. Accordingly, this constitutive law is only invariant by diffeomorphisms preserving $\hat{\omega}$, and this is exactly the definition of the Lie pseudogroup of conformal transformations [13].

□

By chance, the control community has been interested during a while by these new techniques for dealing with PD control theory but mostly restricting to operators with constant coefficients [4, 20, 21]. The following example, coming from partial differential (PD) control theory, will allow the reader to become familiar with these new tools and to understand why they are related to the mathematics of GW. Accordingly, the "relative" parametrization of the Cauchy stress operator has thus nothing to do with the mathematical background of elasticity theory. In particular, the way to simplify the GW equations by bringing them back to the d'Alembert operator while adding a few differential constraints (Compare [14, 15, 22–24]) through a reference to the so-called "gauge invariance" is rather a physical argument and *not* a mathematical one.

Example 1.3. Let us consider the first order operator with two independent variables (x^1, x^2):

$$\mathcal{D}_1 : \left(\eta^1, \eta^2\right) \to d_2\eta^1 - d_1\eta^2 + x^2\eta^2 = \zeta \tag{4}$$

The non-commutative ring of differential operators involved in a formal study is $D = K[d_1, d_2]$ with $K = \mathbb{Q}(x^1, x^2) = \mathbb{Q}(x)$ and the characters of this involutive system are $\left(\alpha_1^1 = 2, \alpha_1^2 = 1\right)$ with $\beta_1^2 = 2 - 1 = 1$ as there is only one equation. Multiplying on the left by a test function λ and integrating by parts, the corresponding adjoint operator is described by:

$$ad(\mathcal{D}_1) : \lambda \to \left(-d_2\lambda = \mu^1, d_1\lambda + x^2\lambda = \mu^2\right)$$

Using crossed derivatives, this operator is injective because $\lambda = d_2\mu^2 + d_1\mu^1 + x^2\mu^1$ and we even obtain a lift $id : \lambda \to \mu \to \lambda$. Substituting, we get the second order involutive operator $ad(\mathcal{D}) : (\mu^1, \mu^2) \to (\nu^1, \nu^2)$ with characters $\left(\alpha_2^1 = 1, \alpha_2^2 = 1\right)$, namely:

$$\left\{\begin{array}{ll} \boxed{d_{22}\mu^2 + d_{12}\mu^1} + x^2 d_2\mu^1 + 2\mu^1 & = \nu^1 \\ \boxed{d_{12}\mu^2 + d_{11}\mu^1} + 2x^2 d_1\mu^1 + x^2 d_2\mu^2 + \left(x^2\right)^2\mu^1 - \mu^2 & = \nu^2 \end{array}\right. \quad \begin{array}{|cc|} 1 & 2 \\ 1 & \bullet \\ \hline \end{array}$$

allowing to define a second order operator $\mathcal{D}$ by using the fact that $ad(ad(\mathcal{D})) = \mathcal{D}$. This operator is involutive and the only corresponding generating CC is $d_2\nu^2 - d_1\nu^1 - x^2\nu^1 = 0$. Therefore, ν^2 is differentially dependent on ν^1 but ν^1 is also differentially dependent on ν^2. Multiplying on the left by a test function θ and integrating by parts, the corresponding adjoint operator is:

$$\mathcal{D}_{-1} : \theta \to \left(d_1\theta - x^2\theta = \xi^1, -d_2\theta = \xi^2\right)$$

Multiplying now the first equation of $ad(\mathcal{D})$ by the test function ξ^1, the second equation by the test function ξ^2, adding and integrating by parts, we get the second order operator:

$$\mathcal{D} : \left(\xi^1, \xi^2\right) \to \left\{\begin{array}{ll} \boxed{d_{22}\xi^1 + d_{12}\xi^2} - x^2 d_2\xi^2 - 2\xi^2 & = \eta^2 \\ \boxed{d_{12}\xi^1 + d_{11}\xi^2} - x^2 d_2\xi^1 - 2x^2 d_1\xi^2 + \xi^1 + \left(x^2\right)^2\xi^2 & = \eta^1 \end{array}\right. \quad \begin{array}{|cc|} 1 & 2 \\ 1 & \bullet \\ \hline \end{array} \tag{5}$$

which is easily seen to be a parametrization of $\mathcal{D}_1$. This operator is involutive and the kernel of this parametrization has differential rank equal to 1 because ξ^1 or ξ^2 can be given arbitrarily.

As we are using the Lagrange multiplier $\left(\xi^1, \xi^2\right)$, we consider in fact the PD equation $\xi^1\nu^1 + \xi^2\nu^2$. Hence, we could indeed consider each term *separately*, that is using *independently* each equation as we shall see later on, ν^1 for a certain ξ and ν^2 for a certain ξ'. Equivalently (exactly like Morera and Maxwell did as we shall see later on), keeping $\xi^1 = \xi$ while setting $\xi^2 = 0$, we now obtain the *first second order minimal parametrization*

$$\xi \to \left(d_{22}\xi = \eta^2, d_{12}\xi - x^2 d_2\xi + \xi = \eta^1\right) \tag{6}$$

This system is again involutive, and the parametrization is minimal because the kernel of this parametrization has differential rank equal to 0. With a similar comment, setting now $\xi^1 = 0$ while keeping $\xi^2 = \xi'$, we get the *second second order minimal parametrization*:

$$\xi' \rightarrow \left(d_{12}\xi' - x^2 d_2\xi' - 2\xi' = \eta^2, d_{11}\xi' - 2x^2 d_1\xi' + \left(x^2\right)^2 \xi' = \eta^1\right), \qquad (7)$$

which is again easily seen to be involutive by exchanging x^1 with x^2.

With again a similar comment, setting now $\xi^1 = d_1\phi, \xi_2 = -d_2\phi$ in the canonical parametrization, we obtain the *third different second order minimal parametrization*:

$$\phi \rightarrow \left(x^2 d_{22}\phi + 2d_2\phi = \eta^2, x^2 d_{12}\phi - \left(x^2\right)^2 d_2\phi + d_1\phi = \eta^1\right) \qquad (8)$$

We are now ready for understanding the meaning and usefulness of what we have defined and called "*relative parametrization*" in [24] by imposing the *differential constraint* $d_2\xi^1 + d_1\xi^2 = 0$. First of all, *we have to prove that such a constraint is compatible*. For this, taking into account the constraint, we have the following first order system defined over K:

$$\left\{\begin{array}{lcl} d_2\xi^1 + d_1\xi^2 & = & 0 \\ d_2\xi^2 + \dfrac{2}{x^2}\xi^2 & = & -\dfrac{1}{x^2}\eta^2 \\ d_1\xi^2 - x^2\xi^2 - \dfrac{1}{x^2}\xi^1 & = & -\dfrac{1}{x^2}\eta^1 \end{array}\right. \quad \begin{array}{|cc|} 1 & 2 \\ 1 & 2 \\ 1 & \bullet \end{array}$$

We let the reader prove that this first order system is involutive with full class 2 and characters $(\alpha_1^1 = 1, \alpha_1^2 = 0)$ and that the only CC involved is the initial system for η (The reader will discover that this checking is quite harder that what one could believe on such an elementary example).

We obtain therefore the new *first order relative parametrization*:

$$\left(\xi^1, \xi^2\right) \rightarrow \left(-x^2 d_2\xi^2 - 2\xi^2 = \eta^2, x^2 d_2\xi^1 + \left(x^2\right)^2 \xi^2 + \xi^1 = \eta^1\right) \quad mod\left(d_2\xi^1 + d_1\xi^2 = 0\right)$$

In a different way, we may add the differential constraint $d_1\xi^1 + d_2\xi^2 = 0$ but we have to check similarly that it is compatible with the previous parametrization. For this, we have to consider the following second order involutive system with five equations, which are easily seen to be involutive, obtained by adding the constraint *and* its two derivatives to the system like before:

$$\left\{\begin{array}{lcl} d_{22}\xi^2 + d_{12}\xi^1 & = & 0 \\ d_{22}\xi^1 + d_{12}\xi^2 - x^2 d_2\xi^2 - 2\xi^2 & = & \eta^2 \\ d_{12}\xi^2 + d_{11}\xi^1 & = & 0 \\ d_{12}\xi^1 + d_{11}\xi^2 - x^2 d_2\xi^1 - 2x^2 d_1\xi^2 + \xi^1 + \left(x^2\right)^2\xi^2 & = & \eta^1 \\ d_2\xi^2 + d_1\xi^1 & = & 0 \end{array}\right. \begin{array}{|cc|} 1 & 2 \\ 1 & 2 \\ 1 & \bullet \\ 1 & \bullet \\ \bullet & \bullet \end{array}$$

The four generating CC only produce the desired system for (η^1, η^2) as we wished.

We could not impose the condition $\mathcal{D}_{-1}\theta = \xi$ already found as it should give the identity $0 = \eta$.

It is, however, also important to notice that the long strictly exact sequence [25]:

$$0 \rightarrow \theta \overset{\mathcal{D}_{-1}}{\rightarrow} \xi \overset{\mathcal{D}}{\rightarrow} \eta \overset{\mathcal{D}_1}{\rightarrow} \zeta \rightarrow 0$$

splits because we have a lift $id : \zeta \rightarrow (-\partial_1\zeta + x^2\zeta = \eta^1, -\partial_2\eta^2 = \eta^2) \rightarrow \partial_2\eta^1 - \partial_1\eta^2 + x^2\eta^2 = \zeta$.

All the differential modules defined from the operators involved are projective, thus torsion-free, and we notice that $\mathcal{D}_1$ is parametrized by $\mathcal{D}$ which is again parametrized by $\mathcal{D}_{-1}$, exactly like *div* is parametrized by *curl* which is again parametrized by *grad* in vector geometry. Needless to say that such an approach has nothing to do with Lorenz gauge invariance in electromagnetism (EM) and we shall arrive to the same conclusion for GW in GR.

Going on along the historical survey 50 years ago, while the author of this paper was working in GR under the leadership of Prof. A. Lichnerowicz, he became familiar with the Lanczos problems. Since that time, he had no wish at all to enter this kind of game as this domain became the private garden of a few persons, each one writing after another one alternatively, claiming to have the full solution. Also, the papers were covered with "computations" involving awful technical formulas, one paper using Gröbner bases, another computer algebra, another Cartan exterior calculus or Janet bases and so on during these 50 years.

Later on, in 2001 and for different reasons, namely control theory as we just explained, being more familiar with differential homological algebra and the so-called "parametrization problem," the way towards the Lanczos problems became easier as follows [18]. In dimension four, the *only* considered by Lanczos, the Lanczos "potential" $L_{ij,k} = -L_{ji,k}$ has $6 \times 4 = 24$ components. As they must be related by the four relations $L_{ij,k} + L_{jk,i} + L_{ki,j} = 0$, we get 20 independent components, namely the number of (second) Bianchi identities. However, who is speaking about "potential" means "parametrization", ... of what?. Here comes the confusion of Lanczos, too much familiar with electromagnetism (EM) while using mainly quadratic Lagrangians with the Riemann tensor in place of the EM field F and the Bianchi identities as differential constraint in the corresponding variational calculus with constraint. The operator to be parametrized was thus the adjoint of the Riemann operator that we called *Beltrami operator* while the parametrizing operator, that we called *Lanczos operator*, was just the adjoint of the *Bianchi operator*, *going now backwards*, that is from right to left in the adjoint sequence of the Killing resolution presented in the abstract. As for the extension to the conformal framework, it is clear that Lanczos did not even know the Weyl tensor when he lectured in France in 1962, invited by Lichnerowicz. Moreover, it is only recently in 2016 that the author of this paper proved that the analogue of the Bianchi identities is made by an operator of order two when $n = 4$, such a result being tested through computer algebra by his former PhD student A. Quadrat (See [12, 13] and arXiv:1603.05030 for more details). Such a construction, based on difficult results of homological algebra, has been missed by Lanczos and all followers as such tools were only available after 1995 through the works of pure mathematicians not interested by applications. Therefore, the main idea is to replace technical formulas by diagram chasing without *any* formula. As a byproduct, we shall understand, *without any computation*, the confusion done between the Cauchy operator, adjoint of the Killing operator, and the Bianchi operator as explained in the abstract. We shall explain its historical origin as the names of many celebrated scientists are involved in this confusion as we also said in the abstract.

The story ended with a strange letter sent by J. Wheeler back to the author with a one dollar bill attached, refusing to admit the negative answer to his parametrization challenge and claiming that future quantum GR should find a positive answer (!). As a byproduct, the impossibility to parametrize Einstein equations in vacuum can only be found in books of control theory [20, 26].

After this rather historical introduction, the content of the paper is clear:

The second section presents the mathematical tools that are absolutely needed while the third section is dealing with the solution of the parametrization problem. The applications to Einstein equations and the corresponding GW equations is finally presented in the fourth section before concluding. In any case, we want to point out that no one of these methods have ever been used in GR, in particular for the study of GW.

2. Mathematical tools

We start recalling the basic tools from the formal theory of systems of ordinary differential (OD) or partial differential (PD) equations and differential modules needed in order to understand and solve the parametrization problem presented in the abstract. Then we provide the example of the system of infinitesimal lie equations defining contact transformations and conclude the paper with the general parametrization problem existing in continuum mechanics and general relativity for an arbitrary dimension of the ground manifold. As these new tools are difficult and not so well known as we already said, we advise the interested reader to follow them step by step on the explicit motivating examples illustrating this paper, while referring to [4, 12, 20, 22, 25, 27–30] for more details, even though the paper is rather self-contained and uses standard notations. Parts of the present paper have already been published independently with slight differences in [2, 12, 13, 18, 20–24, 31] but the present paper is mainly revisiting the mathematical foundations of GW in the framework of differential homological algebra.

2.1 System theory

If X is a manifold of dimension n with local coordinates $(x) = (x^1, \dots, x^n)$, we denote as usual by $T = T(X)$ the *tangent bundle* of X, by $T^* = T^*(X)$ the *cotangent bundle*, by $\wedge^r T^*$ the *bundle of r-forms* and by $S_q T^*$ the *bundle of q-symmetric tensors*. More generally, let E be a *vector bundle* over X with local coordinates (x^i, y^k) for $i = 1, \dots, n$ and $k = 1, \dots, m$ simply denoted by (x, y), *projection* $\pi : E \to X : (x, y) \to (x)$ and changes of local coordinate $\bar{x} = \varphi(x), \bar{y} = A(x)y$. We shall denote by E^* the vector bundle obtained by inverting the matrix A of the changes of coordinates, exactly like T^* is obtained from T. We denote by $f : X \to E : (x) \to (x, y = f(x))$ a global *section* of E, that is a map such that $\pi \circ f = id_X$ but local sections over an open set $U \subset X$ may also be considered when needed. Under a change of coordinates, a section transforms like $\bar{f}(\varphi(x)) = A(x) f(x)$ and the changes of the derivatives can also be obtained with more work. We shall denote by $J_q(E)$ the *q-jet bundle* of E with local coordinates $\left(x^i, y^k, y^k_i, y^k_{ij}, \dots\right) = \left(x, y_q\right)$ called *jet coordinates* and sections $f_q : (x) \to \left(x, f^k(x), f^k_i(x), f^k_{ij}(x), \dots\right) = \left(x, f_q(x)\right)$ transforming like the sections $j_q(f) : (x) \to \left(x, f^k(x), \partial_i f^k(x), \partial_{ij} f^k(x), \dots\right) = \left(x, j_q(f)(x)\right)$ where both f_q and $j_q(f)$ are over the section f of E. For any $q \geq 0$, $J_q(E)$ is a vector bundle over X with projection π_q, while $J_{q+r}(E)$ is a vector bundle over $J_q(E)$ with projection $\pi^{q+r}_q, \forall r \geq 0$.

Definition 2.A.1: A *linear system* of order q on E is a vector sub-bundle $R_q \subset J_q(E)$ and a *solution* of R_q is a section f of E such that $j_q(f)$ is a section of R_q. With a slight abuse of language, the set of local solutions will be denoted by $\Theta \subset E$.

Let $\mu = (\mu_1, \dots, \mu_n)$ be a multi-index with *length* $|\mu| = \mu_1 + \dots + \mu_n$, *class* i if $\mu_1 = \dots = \mu_{i-1} = 0, \mu_i \neq 0$ and $\mu + 1_i = (\mu_1, \dots, \mu_{i-1}, \mu_i + 1, \mu_{i+1}, \dots, \mu_n)$. We set $y_q = \{y^k_\mu | 1 \leq k \leq m, 0 \leq |\mu| \leq q\}$ with $y^k_\mu = y^k$ when $|\mu| = 0$. If E is a vector bundle over X and $J_q(E)$ is the *q-jet bundle* of E, then both sections $f_q \in J_q(E)$ and $j_q(f) \in J_q(E)$ are over the section $f \in E$. There is a natural way to distinguish them by introducing the *Spencer* operator $d : J_{q+1}(E) \to T^* \otimes J_q(E)$ with components $\left(df_{q+1}\right)^k_{\mu,i}(x) = \partial_i f^k_\mu(x) - f^k_{\mu+1_i}(x)$. The kernel of d consists of sections such that $f_{q+1} = j_1(f_q) = j_2(f_{q-1}) = \dots = j_{q+1}(f)$. Finally, if $R_q \subset J_q(E)$ is a *system* of order q on E locally defined by linear equations $\Phi^\tau(x, y_q) \equiv a^{\tau\mu}_k(x) y^k_\mu = 0$ and local coordinates (x, z) for the parametric jets up to order q, the *r-prolongation* $R_{q+r} = \rho_r(R_q) = J_r(R_q) \cap J_{q+r}(E) \subset J_r(J_q(E))$ is locally defined when $r = 1$ by the linear equations $\Phi^\tau(x, y_q) = 0, d_i \Phi^\tau(x, y_{q+1}) \equiv a^{\tau\mu}_k(x) y^k_{\mu+1_i} + \partial_i a^{\tau\mu}_k(x) y^k_\mu = 0$ and has *symbol* $g_{q+r} = R_{q+r} \cap S_{q+r} T^* \otimes E \subset J_{q+r}(E)$ if one looks at the *top order terms*. If $f_{q+1} \in R_{q+1}$ is over $f_q \in R_q$, differentiating the identity $a^{\tau\mu}_k(x) f^k_\mu(x) \equiv 0$ with respect to x^i and substracting the identity $a^{\tau\mu}_k(x) f^k_{\mu+1_i}(x) + \partial_i a^{\tau\mu}_k(x) f^k_\mu(x) \equiv 0$, we obtain the identity $a^{\tau\mu}_k(x)\left(\partial_i f^k_\mu(x) - f^k_{\mu+1_i}(x)\right) \equiv 0$ and thus the restriction $d : R_{q+1} \to T^* \otimes R_q$. More generally, we have the restriction:

$$\boxed{d : \wedge^s T^* \otimes R_{q+1} \to \wedge^{s+1} T^* \otimes R_q : \left(f^k_{\mu,I}(x) dx^I\right) \to \left(\left(\partial_i f^k_{\mu,I}(x) - f^k_{\mu+1_i,I}(x)\right) dx^i \wedge dx^I\right)} \tag{9}$$

with standard multi-index notation for exterior forms and one can easily check that $d \circ d = 0$. The restriction of $-d$ to the symbol is called the *Spencer* map δ in the sequences:

$$\boxed{\wedge^{s-1} T^* \otimes g_{q+r+1} \xrightarrow{\delta} \wedge^s T^* \otimes g_{q+r} \xrightarrow{\delta} \wedge^{s+1} T^* \otimes g_{q+r-1}} \tag{10}$$

because $\delta \circ \delta = 0$ similarly, leading to the purely algebraic *δ-cohomology* $H^s_{q+r}(g_q)$ at $\wedge^s T^* \otimes g_{q+r}$ (See [4, 6, 7, 12, 13, 25, 27–29, 31] for details and examples).

Definition 2.A.2: A system R_q is said to be *formally integrable* (FI) when all the equations of order $q + r$ are obtained by r prolongations *only*, $\forall r \geq 0$ or, equivalently, when the projections $\pi^{q+r+s}_{q+r} : R_{q+r+s} \to R^{(s)}_{q+r} \subseteq R_{q+r}$ are such that $R^{(s)}_{q+r} = R_{q+r}, \forall r, s \geq 0$.

Finding an intrinsic test has been achieved by D.C. Spencer in 1965–1970 [6, 7] along coordinate dependent lines sketched by M. Janet in 1920 [20, 25, 29]. The next procedure providing a *Pommaret basis* and where *one may have to change linearly the independent variables if necessary*, is intrinsic though it must be checked in a particular coordinate system called *δ-regular* [4, 25, 27–29].

- *Equations of class n*: Solve the maximum number β_q^n of equations with respect to the jets of order q and class n. Then, call $(x^1, \dots, x^n)$ *multiplicative variables*.

- *Equations of class* $i \geq 1$: Solve the maximum number β_q^i of *remaining* equations with respect to the jets of order q and class i. Then, call $(x^1, \dots, x^i)$ *multiplicative variables* and $(x^{i+1}, \dots, x^n)$ *non-multiplicative variables*.

- *Remaining equations equations of order* $\leq q-1$: Call $(x^1, \dots, x^n)$ *non-multiplicative variables*.

In actual practice, we shall use a *Janet tabular* where the multiplicative "variables" are in upper left position while the non-multiplicative variables are represented by dots in lower right position.

Definition 2.A.3: A system of PD equations is said to be *involutive* if its first prolongation can be obtained by prolonging its equations only with respect to the corresponding multiplicative variables. In that case, we may introduce the *characters* $\alpha_q^i = m\frac{(q+n-i-1)!}{(q-1)!((n-i)!} - \beta_q^i$ for $i = 1, \dots, n$ with $\alpha_q^1 \geq \dots \geq \alpha_q^n \geq 0$ and we have $dim\left(g_q\right) = \alpha_q^1 + \dots + \alpha_q^n$ while $dim\left(g_{q+1}\right) = \alpha_q^1 + \dots + n\alpha_q^n$.

Remark 2.A.4: As long as the *prolongation/projection* (PP) procedure allowing to find two integers $r, s \geq 0$ such that the system $R_{q+r}^{(s)}$ is involutive, has not been achieved, *nothing* can be said about the CC (Fine examples can be found in [12, 16]).

When R_q is involutive, the operator $\mathcal{D} : E \xrightarrow{j_q} J_q(E) \xrightarrow{\Phi} J_q(E)/R_q = F_0$ of order q is said to be *involutive*. Introducing the *Janet bundles* $F_r = \wedge^r T^* \otimes J_q(E)/\left(\wedge^r T^* \otimes R_q + \delta\left(S_{q+1}T^* \otimes E\right)\right)$, we obtain the *linear Janet sequence* (Introduced in [25]):

$$0 \rightarrow \Theta \rightarrow E \underset{q}{\overset{\mathcal{D}}{\rightarrow}} F_0 \underset{1}{\overset{\mathcal{D}_1}{\rightarrow}} F_1 \underset{1}{\overset{\mathcal{D}_2}{\rightarrow}} \dots \underset{1}{\overset{\mathcal{D}_n}{\rightarrow}} F_n \rightarrow 0 \tag{11}$$

where each other operator is first order involutive and generates the CC of the preceding one.

Similarly, introducing the *Spencer bundles* $C_r = \wedge^r T^* \otimes R_q/\delta\left(\wedge^{r-1}T^* \otimes g_{q+1}\right)$, we obtain the *linear Spencer sequence* induced by the Spencer operator, that can be linked to the Janet sequence [25, 29]:

$$0 \rightarrow \Theta \overset{j_q}{\rightarrow} C_0 \underset{1}{\overset{D_1}{\rightarrow}} C_1 \underset{1}{\overset{D_2}{\rightarrow}} \dots \underset{1}{\overset{D_n}{\rightarrow}} C_n \rightarrow 0 \tag{12}$$

It must be noticed, as we shall see in Section 4, that the Killing operator/system is FI if and only if the metric ω has constant Riemanniann curvature (for example the flat Minkowski metric) but is *not* involutive and the Janet sequence cannot be exhibited. As for the conformal Killing operator/system obtained by eliminating the arbitrary function $A(x)$ in the inhomogeneous system $\mathcal{L}(\xi)\omega = A(x)\omega$ or by considering simply the system $\mathcal{L}(\xi)\hat{\omega} = 0$, it is FI if and only if ω has vanishing Weyl tensor but is *not* involutive and the order of the successive CC operators may change with the ground dimension n, the worst situation being for $n = 4$ as the analogue of the Bianchi operator is now an operator of order two indeed [12, 13]. As these results highly depend on the Spencer δ-cohomology, it is clear that they are neither known nor acknowledged and it follows that the mathematical foundations of conformal geometry must be entirely revisited.

2.2 Module theory

Let K be a *differential field* with n commuting derivations $(\partial_1, \ldots, \partial_n)$ and consider the ring $D = K[d_1, \ldots, d_n] = K[d]$ of differential operators with coefficients in K with n commuting formal derivatives satisfying $d_i a = a d_i + \partial_i a$ in the operator sense. If $P = a^\mu d_\mu \in D = K[d]$, the highest value of $|\mu|$ with $a^\mu \neq 0$ is called the *order* of the *operator* P and the ring D with multiplication $(P, Q) \to P \circ Q = PQ$ is filtred by the order q of the operators. We have the *filtration* $0 \subset K = D_0 \subset D_1 \subset \ldots \subset D_q \subset \ldots \subset D_\infty = D$. As an algebra, D is generated by $K = D_0$ and $T = D_1/D_0$ with $D_1 = K \oplus T$ if we identify an element $\xi = \xi^i d_i \in T$ with the vector field $\xi = \xi^i(x)\partial_i$ of differential geometry, but with $\xi^i \in K$ now. It follows that $D = {}_D D_D$ is a *bimodule* over itself, being at the same time a left D-module by the composition $P \to QP$ and a right D-module by the composition $P \to PQ$. We define the *adjoint* functor $ad : D \to D^{op} : P = a^\mu d_\mu \to ad(P) = (-1)^{|\mu|} d_\mu a^\mu$ and we have $ad(ad(P)) = P$ both with $ad(PQ) = ad(Q)ad(P), \forall P, Q \in D$. It follows that any operator can be considered as the adjoint of its own adjoint. Such a definition can be extended by linearity to any matrix of operators by using the transposed matrix of adjoint operators (See [4, 5, 11–13, 18, 20, 21, 24] for more details and applications to control theory or mathematical physics).

Accordingly, if $y = (y^1, \ldots, y^m)$ are differential indeterminates, then D acts on y^k by setting $d_i y^k = y^k_i \to d_\mu y^k = y^k_\mu$ with $d_i y^k_\mu = y^k_{\mu+1_i}$ and $y^k_0 = y^k$. We may therefore use the jet coordinates in a formal way as in the previous section. Therefore, if a system of OD/PD equations is written in the form $\Phi^\tau \equiv a^{\tau\mu}_k y^k_\mu = 0$ with coefficients $a \in K$, we may introduce the free differential module $Dy = Dy^1 + \ldots + Dy^m \simeq D^m$ and consider the differential *module of equations* $I = D\Phi \subset Dy$, both with the residual *differential module* $M = Dy/D\Phi$ or D-module and we may set $M = {}_D M$ if we want to specify the ring of differential operators. We may introduce the formal *prolongation* with respect to d_i by setting $d_i \Phi^\tau \equiv a^{\tau\mu}_k y^k_{\mu+1_i} + (\partial_i a^{\tau\mu}_k) y^k_\mu$ in order to induce maps $d_i : M \to M : \bar{y}^k_\mu \to \bar{y}^k_{\mu+1_i}$ by residue with respect to I if we use to denote the residue $Dy \to M : y^k \to \bar{y}^k$ by a bar like in algebraic geometry. However, for simplicity, we shall not write down the bar when the background will indicate clearly if we are in Dy or in M. As a byproduct, the differential modules we shall consider will always be *finitely generated* $(k = 1, \ldots, m < \infty)$ and *finitely presented* $(\tau = 1, \ldots, p < \infty)$. Equivalently, introducing the *matrix of operators* $\mathcal{D} = (a^{\tau\mu}_k d_\mu)$ with m columns and p rows, we may introduce the morphism $D^p \xrightarrow{\mathcal{D}} D^m : (P_\tau) \to (P_\tau \Phi^\tau)$ over D by acting with D *on the left of these row vectors* while acting with $\mathcal{D}$ *on the right of these row vectors* by composition of operators with $im(\mathcal{D}) = I$. The *presentation* of M is defined by the exact cokernel sequence $D^p \xrightarrow{\mathcal{D}} D^m \to M \to 0$. We notice that the presentation only depends on K, D and Φ or $\mathcal{D}$, that is to say never refers to the concept of (explicit local or formal) solutions. It follows from its definition that M can be endowed with a *quotient filtration* obtained from that of D^m which is defined by the order of the jet coordinates y_q in $D_q y$. We have therefore the *inductive limit* $0 \subseteq M_0 \subseteq M_1 \subseteq \ldots \subseteq M_q \subseteq \ldots \subseteq M_\infty = M$ with $d_i M_q \subseteq M_{q+1}$

and $M = DM_q$ for $q \gg 0$ with prolongations $D_r M_q \subseteq M_{q+r}, \forall q, r \geq 0$. It may be sometimes quite difficult to work out I_q or M_q from a given presentation which is not involutive [4].

Definition 2.B.1: An exact sequence of morphisms finishing at M is said to be a *resolution* of M. If the differential modules involved apart from M are free, that is isomorphic to a certain power of D, we shall say that we have a *free resolution* of M.

Having in mind that K is a left D-module with the action $(D,K) \to K : (d_i, a) \to \partial_i a$ and that D is a bimodule over itself with $PQ \neq QP$, *we have only two possible constructions*:

Definition 2.B.2: We may define the right (*care*) differential module $hom_D(M,D)$ with $(fP)(m) = (f(m))P \Rightarrow (fPQ)(m) = ((fP)(m))Q = ((f(m))P)Q = (f(m))PQ$.

Definition 2.B.3: We define the *system* $R = hom_K(M,K)$ and set $R_q = hom_K(M_q, K)$ as the *system of order* q. We have the *projective limit* $R = R_\infty \to \ldots \to R_q \to \ldots \to R_1 \to R_0$. It follows that $f_q \in R_q : y^k_\mu \to f^k_\mu \in K$ with $a^{\tau\mu}_k f^k_\mu = 0$ defines a *section at order* q, and we may set $f_\infty = f \in R$ for a *section* of R. For an arbitrary differential field K, *such a definition has nothing to do with the concept of a formal power series solution* (*care*).

Proposition 2.B.4: When M is a left D-module, then R is also a left D-module.

Proof: As D is generated by K and T as we already said, let us define:

$$(af)(m) = af(m) = f(am), \quad \forall a \in K, \forall m \in M$$
$$(\xi f)(m) = \xi f(m) - f(\xi m), \quad \forall \xi = a^i d_i \in T, \forall m \in M$$

In the operator sense, it is easy to check that $d_i a = a d_i + \partial_i a$ and that $\xi\eta - \eta\xi = [\xi, \eta]$ is the standard bracket of vector fields. We finally get $(d_i f)^k_\mu = (d_i f)\left(y^k_\mu\right) = \partial_i f^k_\mu - f^k_{\mu+1_i}$ and thus recover *exactly* the Spencer operator of the previous section though *this is not evident at all*. We also get $(d_i d_j f)^k_\mu = \partial_{ij} f^k_\mu - \partial_i f^k_{\mu+1_j} - \partial_j f^k_{\mu+1_i} + f^k_{\mu+1_i+1_j} \Rightarrow d_i d_j = d_j d_i, \forall i,j = 1, \ldots, n$ and thus $d_i R_{q+1} \subseteq R_q \Rightarrow d_i R \subset R$ induces a well defined operator $R \to T^* \otimes R : f \to dx^i \otimes d_i f$. This operator has been first introduced, up to sign, by F. S. Macaulay as early as in 1916 but this is still not ackowledged. For more details on the Spencer operator and its applications, the reader may look at [12, 13, 25, 27–30].

□

Definition 2.B.5: With any differential module M, we shall associate the *graded module* $G = gr(M)$ over the polynomial ring $gr(D) \simeq K[\chi]$ by setting $G = \oplus_{q=0}^{\infty} G_q$ with $G_q = M_q / M_{q-1}$, and we get $g_q = G_q^*$ where the *symbol* g_q is defined by the short exact sequences:

$$0 \to M_{q-1} \to M_q \to G_q \to 0 \quad \Leftrightarrow \quad 0 \to g_q \to R_q \to R_{q-1} \to 0$$

We have the short exact sequences $0 \to D_{q-1} \to D_q \to S_q T \to 0$ leading to $gr_q(D) \simeq S_q T$ and we may set as usual $T^* = hom_K(T,K)$ in a coherent way with differential geometry.

The two following definitions, which are well known in commutative algebra, are also valid (with more work) in the case of differential modules (See [4, 20] for more details or the references [4, 8–10, 12, 13, 29] for an introduction to homological algebra and diagram chasing).

Definition 2.B.6: The set of elements $t(M) = \{m \in M | \exists 0 \neq P \in D, Pm = 0\} \subseteq M$ is a differential module called the *torsion submodule* of M. More generally, a module M is called a *torsion module* if $t(M) = M$ and a *torsion-free module* if $t(M) = 0$. In the short exact sequence $0 \to t(M) \to M \to M' \to 0$, the module M' is torsion-free. Its defining module of equations I' is obtained by adding to I a representative basis of $t(M)$ set up to zero and we have thus $I \subseteq I'$.

Definition 2.B.7: A differential module F is said to be *free* if $F \simeq D^r$ for some integer $r > 0$ and we shall *define* $rk_D(F) = r$. If F is the biggest free differential module contained in M, then M/F is a torsion differential module and $hom_D(M/F, D) = 0$. In that case, we shall *define* the *differential rank* of M to be $rk_D(M) = rk_D(F) = r$. Accordingly, if M is defined by a linear involutive operator of order q, then $rk_D(M) = \alpha_q^n$.

Proposition 2.B.8: If $0 \to M' \to M \to M'' \to 0$ is a short exact sequence of differential modules and maps or operators, we have $rk_D(M) = rk_D(M') + rk_D(M'')$.

In the general situation, let us consider the sequence $M' \xrightarrow{f} M \xrightarrow{g} M''$ of modules which may not be exact. Then, we may define the *coboundary* submodule $B = im(f) \subseteq M$, the *cocycle* submodule $Z = ker(g) \subseteq M$ and the *cohomology* module $H = Z/B$.

Using the last (delicate) proposition, we may provide the following definitions that will be in the heart of the parametrization problem, successively in the operator and module frameworks.

Definition 2.B.9: When $\mathcal{D} = \Phi \circ j_q : E \to F$ is a linear differential operator of order q with coefficients in a differential field K, between the sections of two vector bundles E with $dim(E) = m$ and F with $dim(F) = p$, we shall define in a formal way the *differential rank* $rk_D(\mathcal{D}) = m - \alpha_q^n = \beta_q^n = rk_D(im(\mathcal{D})) \leq p$ by introducing the characters of the corresponding linear system $R_q = ker(\Phi) \subseteq J_q(E)$ of order q over E. We have thus $rk_D(\mathcal{D}) \leq inf\,(m, p)$ (See [4, 20] for details). The order of an operator will be indicated under it arrow.

Definition 2.B.10: When a differential module M is defined by the presentation $D^p \underset{q}{\xrightarrow{\mathcal{D}}} D^m \to M \to 0$, we shall introduce the differential module $I = im(\mathcal{D}) \subseteq D^m$ and set $rk_D(\mathcal{D}) = rk_D(I) = m - rk_D(M)$ in a coherent way with the last proposition and definition.

We obtain the important theorem which is generalizing to operators the rank property of a $m \times p$ matrix, even when $\mathcal{D}$ and $ad(\mathcal{D})$ are neither FI nor involutive ([4, 20, 29], p. 340):

Theorem 2.B.11: One has $rk_D(\mathcal{D}) = rk_D(ad(\mathcal{D})) \leq inf\,(m, p)$.

In order to conclude this section, we may say that the main difficulty met when passing from the differential framework to the algebraic framework is the "*inversion*" of arrows. Indeed, when an operator is injective, that is when we have the exact sequence $0 \to E \xrightarrow{\mathcal{D}} F$ with $dim(E) = m$, $dim(F) = p$, like in the case of the operator $0 \to E \xrightarrow{j_q} J_q(E)$, on the contrary, using differential modules, we have the epimorphism $D^p \xrightarrow{\mathcal{D}} D^m \to 0$. The case of a formally surjective operator, like the *div* operator, described by the exact sequence $E \xrightarrow{\mathcal{D}} F \to 0$ is now providing the exact sequence of differential modules $0 \to D^p \xrightarrow{\mathcal{D}} D^m \to M \to 0$ because $\mathcal{D}$ has no CC.

3. Parametrization problem

In this section, we shall set up and solve the minimum parametrization problem by comparing the differential geometric approach and the differential algebraic approach. In fact, both sides are essential because certain concepts, like "torsion", are simpler in the module approach while others, like "involution" are simpler in the operator approach. However, the reader must never forget that the "*extension modules*" or the "*side changing functor*" are pure product of *differential homological algebra* with no system counterpart. The link between "*differential duality*" and the "*adjoint operator*" may not be evident at all, even for people familiar with mathematical physics [2, 4, 12, 13, 18, 22, 29].

Let us start with a given linear differential operator $\eta \overset{\mathcal{D}_1}{\rightarrow} \zeta$ between the sections of two given vector bundles F_0 and F_1 of respective fiber dimension m and p. Multiplying the equations $\mathcal{D}_1\eta = \zeta$ by p test functions λ considered as a section of the adjoint vector bundle $ad(F_1) = \wedge^n T^* \otimes F_1^*$ and integrating by parts as we did in the introduction, we may introduce the adjoint vector bundle $ad(F_0) = \wedge^n T^* \otimes F_0^*$ with sections μ in order to obtain the adjoint operator $\mu \overset{ad(\mathcal{D}_1)}{\longleftarrow} \lambda$, writing on purpose the arrow *backwards*. More generally, let us consider a differential sequence:

$$\xi \overset{\mathcal{D}}{\rightarrow} \eta \overset{\mathcal{D}_1}{\rightarrow} \zeta \tag{13}$$

such that $\mathcal{D}_1$ generates the CC of $\mathcal{D}$ or, equivalently, such that $\mathcal{D}_1$ is parametrized by $\mathcal{D}$.

We may introduce the adjoint differential sequence:

$$\nu \overset{ad(\mathcal{D})}{\longleftarrow} \mu \overset{ad(\mathcal{D}_1)}{\longleftarrow} \lambda \tag{14}$$

As we have $\mathcal{D}_1 \circ \mathcal{D} = 0$, we obtain $ad(\mathcal{D}) \circ ad(\mathcal{D}_1) = 0$. However, if $\mathcal{D}_1$ generates the CC of $\mathcal{D}$, then $ad(\mathcal{D})$ may not generate the CC of $ad(\mathcal{D}_1)$. Such a situation may not be satisfied as we saw and the so-called *extension modules* have been introduced in order to measure these "*gaps.*"

In order to pass to the differential module framework, let us introduce the free differential modules $D\xi \simeq D^l, D\eta \simeq D^m, D\zeta \simeq D^p$. We have similarly the adjoint free differential modules $D\nu \simeq D^l, D\mu \simeq D^m, D\lambda \simeq D^p$, because $dim(ad(E)) = dim(E)$ and $hom_D(D^m, D) \simeq D^m$. Of course, in actual practice, *the geometric meaning is totally different* because we have volume forms in the dual framework. We have thus obtained the formally exact sequence of differential modules:

$$\begin{array}{ccccccc}
D^p & \overset{\mathcal{D}_1}{\rightarrow} & D^m & & \overset{\mathcal{D}}{\rightarrow} & & D^l \rightarrow M \rightarrow 0 \\
 & & & \searrow & & \nearrow & \\
 & & & & M_1 & & \\
 & & & \nearrow & & \searrow & \\
 & & 0 & & & & 0
\end{array} \tag{15}$$

with $rk_D(\mathcal{D}) + rk_D(\mathcal{D}_1) = m \Leftrightarrow rk_D(M) + rk_D(M_1) = l$. We have the adjoint sequence:

$$D^p \overset{ad(\mathcal{D}_1)}{\leftarrow} D^m \overset{ad(\mathcal{D})}{\leftarrow} D^l$$

with $rk_D(ad(\mathcal{D})) + rk_D(ad(\mathcal{D}_1)) = m$ and we may thus state [4, 9, 10, 20]:

Definition 3.1: We may define the *zero differential extension module* $ext^0(M) = ker(ad(\mathcal{D}))$ and the *first differential extension module* $ext^1(M)$ to be the cohomology of this sequence at D^m. The latter is a torsion module because it has vanishing differential rank. They only depend on M.

Theorem 3.2: There is a constructive test in order to find out whether a differential operator $\mathcal{D}_1$ can be parametrized or not (Example 1.3 or the *div* operator with $n = 3$).

Proof: The test has five steps along with the following diagram in operator language:

$\boxed{1} \Rightarrow$ Start with $\mathcal{D}_1 : \eta \to \zeta$,
$\boxed{2} \Rightarrow$ Construct $ad(\mathcal{D}_1) : \lambda \to \mu$,
$\boxed{3} \Rightarrow$ Construct its CC as an operator $ad(\mathcal{D}) : \mu \to \nu$,
$\boxed{4} \Rightarrow$ Construct its adjoint $\mathcal{D} = ad(ad(\mathcal{D})) : \xi \to \eta$,
$\boxed{5} \Rightarrow$ Construct the CC $\mathcal{D}_1'$ of $\mathcal{D} : \eta \to \zeta'$.

$$\begin{array}{ccccccccc}
 & & & & & & \zeta' & \boxed{5} \\
 & & & & & \overset{\mathcal{D}_1'}{\nearrow} & & \\
\boxed{4} & \xi & \overset{\mathcal{D}}{\longrightarrow} & \eta & \overset{\mathcal{D}_1}{\longrightarrow} & & \zeta & \boxed{1} \\
 & & & & & & & \\
\boxed{3} & \nu & \overset{ad(\mathcal{D})}{\longleftarrow} & \mu & \overset{ad(\mathcal{D}_1)}{\longleftarrow} & & \lambda & \boxed{2}
\end{array}$$

We have $ad(\mathcal{D}) \circ ad(\mathcal{D}_1) = 0 \Rightarrow \mathcal{D}_1 \circ \mathcal{D} = 0$, that is $\mathcal{D}_1$ is surely among the CC of $\mathcal{D}$ but other CC may also exist. The parametrization is thus existing if and only if we may have $\mathcal{D}_1' = \mathcal{D}_1$.

□

Corollary 3.3: Each new CC *eventually* brought by $\mathcal{D}'$ which is not already a differential consequence of $\mathcal{D}_1$ is providing a torsion element of the differential module M_1 determined by $\mathcal{D}_1$.

Example 3.4: If $\mathcal{D} : \xi \to (d_{22}\xi = \eta^2, d_{12}\xi = \eta^1)$ we have $\mathcal{D}_1 = (\eta^1, \eta^2) \to d_1\eta^2 - d_2\eta^1 = \zeta$ and the only first order generating CC of $ad(\mathcal{D}_1) : \lambda \to (d_2\lambda = \mu^1, -d_1\lambda = \mu^2)$ is $d_1\mu^1 + d_2\mu^2 = \nu'$ while $ad(\mathcal{D}) : (\mu^1, \mu^2) \to d_{12}\mu^1 + d_{22}\mu^2 = \nu$ is a second order operator like $\mathcal{D}$. Hence, if we should like to parametrize $ad(\mathcal{D})$, we should successively find $\mathcal{D}$, $\mathcal{D}_1$, $ad(\mathcal{D}_1)$ and finally get the additional first order CC $d_1\mu^1 + d_2\mu^2 = \nu'$ which is such that $\nu = 0 \Rightarrow d_2\nu' = 0$, that is ν' is a torsion element for the system $d_{12}\mu^1 + d_{22}\mu^2 = 0$.

Example 3.5: Many other examples can be found in ordinary differential control theory because it is known that *a linear control system is controllable if and only if it is parametrizable* (See [4, 20, 29] for more details and examples). In our opinion, the simplest one is provided by the double pendulum in which a rigid bar is able to move horizontally with reference position x and has two pendulums attached with respective length l_1 and l_2 making the (small) angles θ_1 and θ_2 with the vertical. The corresponding control system does not depend on the mass of each pendulum and is:

$$d^2x + l_1 d^2\theta^1 + g\theta^1 = 0, \quad d^2x + l_2 d^2\theta^2 + g\theta^2 = 0$$

where g is the gravity. The classical approach is to prove that this control system is controllable if and only if $l_1 \neq l_2$ by using a tedious computation through the standard Kalman test [20]. However, *equivalently, but this way is still not acknowledged by the control community*, the idea is to prove that the corresponding second order operator $ad(\mathcal{D}_1)$ is injective. We let the reader realize the experiment, prove this result as an exercise and apply the previous theorem in order to work out the parametrizing operator $\mathcal{D}$ of order 4, namely:

$$\left\{\begin{array}{rcl} -l_1 l_2 d^4\phi - g(l_1+l_2)d^2\phi - g^2\phi & = & x \\ l_2 d^4\phi + g d^2\phi & = & \theta_1 \\ l_1 d^4\phi + g d^2\phi & = & \theta_2 \end{array}\right.$$

Example 3.6: As a less academic example, the following diagram is proving that *Einstein equations cannot be parametrized* [1, 2]:

$$\begin{array}{ccccccccccc} & & & & 10 & \overset{Riemann}{\longrightarrow} & 20 & \overset{Bianchi}{\longrightarrow} & 20 & & \\ & & & & \| & \nearrow & \downarrow & & \downarrow & & \\ & & 4 & \overset{Killing}{\longrightarrow} & 10 & \overset{Einstein}{\longrightarrow} & 10 & \overset{div}{\longrightarrow} & 4 & \longrightarrow & 0 \\ & & & & & & & & & & \\ 0 & \longrightarrow & 4 & \overset{Cauchy}{\longleftarrow} & 10 & \overset{Einstein}{\longleftarrow} & 10 & & & & \end{array}$$

and that the *Cauchy* and *Killing* operators (*left side*) have strictly nothing to do with the *Bianchi* and *div* operators (*right side*). According to the last corollary, the $20 - 10 = 10$ new CC are generating the torsion submodule of the differential module defined by the Einstein operator. In the last section we shall explain why such a basis of the torsion module is made by the 10 independent components of the Weyl tensor, *a result which is not evident*, leading to the so-called *Lichnerowicz waves* (in France) [12, 13, 22, 32, 33].

In continuum mechanics, the Cauchy stress tensor may not be symmetric in the so-called *Cosserat media* where the *Cauchy stress equations* are replaced by the *Cosserat couple-stress equations* which are nothing else than the adjoint of the first Spencer operator, *totally different* from the third [27, 28, 34, 35]. When $n = 2$, we shall see that the single Airy function has strictly nothing to do with any perturbation of the metric having three components.

Example 3.7: A similar comment can be done for electromagnetism through the exterior derivative as the first set of Maxwell equations can be parametrized by the EM potential 1-form because $dA = F \Rightarrow dF = 0$ while the second set of Maxwell equations (adjoint of this parametrization) can be parametrized by the EM pseudo-potential, a 3 - skew-symmetric contravariant tensor density in $\wedge^4 T^* \otimes \wedge^3 T$ through the adjoint of the exterior derivative $\wedge^2 T^* \overset{d}{\longrightarrow} \wedge^3 T^*$ [12, 13, 27, 28]. These results, which are deeply supporting the conformal origin of EM [31], are also strengthening the comments we shall make in Section 4 on the origin and existence of gravitational waves [13, 22, 23].

As a summarizing comment, we discover that not only it is sometimes not possible to parametrize a linear differential operator but that, whenever it is possible, not only it is not easy to have an idea about the number of potential functions needed but even more difficult to have *any* idea about the order of the parametrizing operator that may be unexpectedly quite high indeed for the double pendulum. In addition, we have provided in [4], examples showing that the case of variable coefficients is even much more difficult than the case of constant coefficients. Moreover, the mathematical tools involved are sometimes not accessible to intuition like this theorem (See [4, 5, 20] for details):

Theorem 3.8: If M is a differential module, we have the exact sequence of differential modules:

$$0 \rightarrow t(M) \rightarrow M \xrightarrow{\varepsilon} hom_D(hom_D(M,D),D) \tag{16}$$

where the map ε is defined by $\varepsilon(m)(f) = f(m), \forall m \in M, f \in hom_D(M,D)$.

Theorem 3.9: When $\mathcal{D}_1$ can be parametrized or, equivalently, when M_1 is torsion-free and can be thus embedded into a free module D^l, we have thus $rk_D(M_1) = l' \leq l$. There is a constructive procedure in order to embed M_1 into $D^{l'}$, that is to obtain a minimum parametrization.

Proof: The procedure with 4 steps is as follows in the operator language (**Example 1.3**):

$\boxed{1} \Rightarrow$ Start with the formally exact *parametrizing sequence* already constructed by differential biduality. We have thus $im(\mathcal{D}) = ker(\mathcal{D}_1)$ and the corresponding differential module M_1 defined by $\mathcal{D}_1$ is torsion-free by assumption.

$\boxed{2} \Rightarrow$ Construct the adjoint sequence which is also formally exact by assumption.

$\boxed{3} \Rightarrow$ Find a *maximum* set of differentially independent CC $ad(\mathcal{D}') : \mu \rightarrow \nu'$ among the generating CC $ad(\mathcal{D}) : \mu \rightarrow \nu$ of $ad(\mathcal{D}_1)$ in such a way that $im(ad(\mathcal{D}'))$ is a maximum free differential submodule of $im(ad(\mathcal{D}))$ that is any element in $im(ad(\mathcal{D}))$ is differentially algebraic over $im(ad(\mathcal{D}'))$.

$\boxed{4} \Rightarrow$ Using differential duality, construct $\mathcal{D}' = ad(ad(\mathcal{D}'))$.

Let us prove that $\mathcal{D}_1$ generates the CC of $\mathcal{D}'$ in the following double diagram:

$$\begin{array}{ccccccc}
\boxed{4} & & \xi' & & & & \\
 & & \uparrow & \overset{\mathcal{D}'}{\searrow} & & & \\
 & & \xi & \overset{\mathcal{D}}{\longrightarrow} & \eta & \overset{\mathcal{D}_1}{\longrightarrow} & \zeta \quad \boxed{1} \\
 & & & & & & \\
 & & \nu & \overset{ad(\mathcal{D})}{\longleftarrow} & \mu & \overset{ad(\mathcal{D}_1)}{\longleftarrow} & \lambda \quad \boxed{2} \\
 & & \uparrow & \underset{}{\overset{ad(\mathcal{D}')}{\swarrow}} & & & \\
\boxed{3} & & \nu' & & & & \\
 & \swarrow & \uparrow & & & & \\
0 & & 0 & & & &
\end{array}$$

First of all, we have by construction $im(ad(\mathcal{D}')) = im(ad(\mathcal{D}))$ in the bottom diagram and thus:

$$rk_D(ad(\mathcal{D}')) + rk_D(ad(\mathcal{D}_1)) = rk_D(ad(\mathcal{D})) + rk_D(ad(\mathcal{D}_1)) = m$$

Passing to the upper diagram, we have, therefore:

$$rk_D(\mathcal{D}') + rk_D(\mathcal{D}_1) = rk_D(\mathcal{D}) + rk_D(\mathcal{D}_1) = m$$

We have $ad(\mathcal{D}') \circ ad(\mathcal{D}_1) = ad(\mathcal{D}_1 \circ \mathcal{D}') = 0 \Rightarrow \mathcal{D}_1 \circ \mathcal{D}' = 0$ and $\mathcal{D}_1$ is surely *among* the CC of $\mathcal{D}'$. Therefore, the differential sequence $\xi' \overset{\mathcal{D}'}{\longrightarrow} \eta \overset{\mathcal{D}_1}{\longrightarrow} \zeta$ on the operator level or the sequence $D^p \overset{\mathcal{D}_1}{\longrightarrow} D^m \overset{\mathcal{D}'}{\longrightarrow} D^{l'}$ on the differential module level may not be exact. In the latter, we have now $B = im(\mathcal{D}_1) = ker(\mathcal{D}) \subseteq ker(\mathcal{D}') = Z \subseteq D^m$. But we have also $rk_D(B) = m - rk_D(\mathcal{D})$, $rk(Z) = m - rk_D(\mathcal{D}') \Rightarrow rk_D(H) = rk_D(\mathcal{D}) - rk_D(\mathcal{D}') = 0$. Using the fact that $M_1 = coker(\mathcal{D}_1)$ while setting $M_1' = im(\mathcal{D}') \subseteq D^{l'}$, we get the commutative and exact diagram:

$$\begin{array}{ccccccccc}
 & & 0 & & 0 & & & & \\
 & & \downarrow & & \downarrow & & & & \\
0 & \rightarrow & B & \rightarrow & D^m & \rightarrow & M_1 & \rightarrow & 0 \\
 & & \downarrow & & \parallel & & \downarrow & & \\
0 & \rightarrow & Z & \rightarrow & D^m & \rightarrow & M_1' & \rightarrow & 0 \\
 & & \downarrow & & \downarrow & & \downarrow & & \\
 & & H & & 0 & & 0 & & \\
 & & \downarrow & & & & & & \\
 & & 0 & & & & & &
\end{array}$$

A snake chase shows that the kernel of the induced epimorphism $M_1 \rightarrow M_1' \rightarrow 0$ is isomorphic to H and is thus a torsion module because we have just proved that $rk_D(H) = 0$. However, we know that M_1 is a torsion-free module and we reach a contradiction unless $H = 0 \Leftrightarrow M_1 \simeq M_1'$.

In actual practice, as shown in the contact case below, things are not so simple and we shall use the following commutative and exact diagram of differential modules:

$$\begin{array}{ccccccc}
0 \rightarrow ker(ad(\mathcal{D})) \rightarrow & D^l & & \overset{ad(\mathcal{D})}{\longrightarrow} & & D^m & \rightarrow coker(ad(\mathcal{D})) \rightarrow 0 \\
 & & \searrow & & \nearrow & & \\
 & & & L & & & \\
 & & \nearrow & \uparrow & \searrow & & \\
 & 0 & & D^{l'} & & 0 & \\
 & & & \uparrow & & & \\
 & & & 0 & & &
\end{array}$$

Setting $L = D^l / ker(ad(\mathcal{D}))$ and introducing the biggest free differential module $D^{l'} \subseteq L$ we have $l' = rk_D\left(D^{l'}\right) = rk_D(L) \leq rk_D\left(D^l\right) \Rightarrow l' \leq l$, we may *define* the injective (*care*) operator $ad(\mathcal{D}')$ by the composition of monomorphisms $D^{l'} \rightarrow L \rightarrow D^m$ where the second is obtained by picking a basis of $D^{l'}$, lifting it to D^l and pushing it to D^m by applying $ad(\mathcal{D})$. We notice that L can be viewed as the differential module defined by the generating CC of $ad(\mathcal{D})$.

□

Example 3.10: *Contact transformations.*

With $m = n = 3, K = \mathbb{Q}(x^1, x^2, x^3) = \mathbb{Q}(x)$, we may introduce the so-called *contact* 1-form $\alpha = dx^1 - x^3 dx^2$. The system of infinitesimal Lie equations defining the infinitesimal contact transformations is obtained by eliminating the factor $\rho(x)$ in the equations $\mathcal{L}(\xi)\alpha = \rho\alpha$ where $\mathcal{L}$ is the standard Lie derivative. This system is thus only generated by η^1 and η^2 below but is not involutive and one has to introduce η^3 defined by the first order CC operator:

$$(\eta^1, \eta^2, \eta^3) \rightarrow (d_3\eta^1 - d_2\eta^2 - x^3 d_1\eta^2 + \eta^3 = \zeta)$$

in order to obtain the following involutive system with two equations of class 3 and one equation of class 2, a result leading to $\beta_1^3 = 2, \beta_1^2 = 1, \beta_1^1 = 0$ and the characters $\alpha_1^3 = 3 - 2 = 1 < \alpha_1^2 = 3 - 1 = 2, \alpha_1^1 = 3 - 0 = 3$ with sum equal to $1 + 2 + 3 = 6 = dim(g_1) = 3 \times 3 - 3$.

$$\begin{cases} d_3\xi^3 + d_2\xi^2 + 2x^3 d_1\xi^2 - d_1\xi^1 & = & \eta^3 \\ d_3\xi^1 - x^3 d_3\xi^2 & = & \eta^2 \\ d_2\xi^1 - x^3 d_2\xi^2 + x^3 d_1\xi^1 - (x^3)^2 d_1\xi^2 - \xi^3 & = & \eta^1 \end{cases} \quad \begin{array}{|ccc|} 1 & 2 & 3 \\ 1 & 2 & 3 \\ 1 & 2 & \bullet \end{array}$$

The characters are thus $\alpha_1^3 = 3 - 2 = 1 < \alpha_1^2 = 3 - 1 = 2, \alpha_1^1 = 3 - 0 = 3$ with sum equal to $1 + 2 + 3 = 6 = dim(g_1) = 3 \times 3 - 3 = 6$ and we get $dim(g_2) = 3 + (2 \times 2) + (3 \times 1) = 10 = dim(S_2T^* \otimes T) - 8$ along the Janet tabular [20].

In this situation, if M is the differential module defined by this system or the corresponding operator $\mathcal{D}$, we know that $rk_D(M) = \alpha_1^3 = 1 = 3 - 2 = rk_D(D\xi) - rk_D(\mathcal{D})$. Of course, a differential transcendence basis for $\mathcal{D}$ can be the operator $\mathcal{D}' : \xi \rightarrow \{\eta^2, \eta^3\}$ but, in view of the CC, we may equally choose any couple among $\{\eta^1, \eta^2, \eta^3\}$ and we obtain $rk_D(\mathcal{D}') = rk_D(\mathcal{D}) = 2$ in any case, but now $\mathcal{D}'$ is formally surjective, contrary to $\mathcal{D}$. The same result can also be obtained directly from the unique CC or the corresponding operator $\mathcal{D}_1$ defining the differential module M_1. Finally, we have $rk_D(M_1) = 3 - 1 = 2 = rk_D(D\eta) - rk_D(\mathcal{D}_1)$ and we check that we have indeed $rk_D(M) + rk_D(M_1) = 1 + 2 = 3 = rk_D(D\xi)$.

It is well known that such a system can be parametrized by the injective parametrization (See [19, 29] for more details and the study of the general dimension $n = 2p + 1$):

$$-x^3 d_3\phi + \phi = \xi^1, \quad -d_3\phi = \xi^2, \ d_2\phi + x^3 d_1\phi = \xi^3 \ \Rightarrow \ \xi^1 - x^3\xi^2 = \phi$$

Noticing that $\mathcal{D}$ is generated by $\mathcal{D}' : \xi \rightarrow (\eta^1, \eta^2)$, we obtain the operator $ad(\mathcal{D}')$:

$$-d_3\mu^2 - d_2\mu^1 - x^3 d_1\mu^1 = \nu^1, \ x^3 d_3\mu^2 + \mu^2 + x^3 d_2\mu^1 + (x^3)^2 d_1\mu^1 = \nu^2, \ \ -\mu^1 = \nu^3$$

It follows that $\mu^1 = -\nu^3, \mu^2 = \nu^2 + x^3\nu^1$ and, substituing, the only CC:

$$(x^3 d_3\nu^1) + (d_3\nu^2) + (d_2\nu^3 + x^3 d_1\nu^3) = 0$$

which is *exactly* $ad(\mathcal{D}_{-1})$ as can be seen by multiplying by a test function ϕ and integrating by parts. No such computations can be found in the literature on contact structures.

The associated differential sequence is:

$$\begin{array}{|ccccccccc|} 0 \rightarrow & \phi & \stackrel{\mathcal{D}_{-1}}{\rightarrow} & \xi & \stackrel{\mathcal{D}}{\rightarrow} & \eta & \stackrel{\mathcal{D}_1}{\rightarrow} & \zeta & \rightarrow 0 \\ \hline 0 \rightarrow & 1 & \rightarrow & 3 & \rightarrow & 3 & \rightarrow & 1 & \rightarrow 0 \end{array}$$

with Euler-Poincaré characteristic $1 - 3 + 3 - 1 = 0$ but is *not* a Janet sequence because $\mathcal{D}_{-1}$ is not involutive, its completion to involution being the trivially involutive operator $j_1 : \phi \to j_1(\phi)$.

Introducing the ring $D = K[d_1, d_2, d_3] = K[d]$ of linear differential operators with coefficients in the differential field K, the corresponding differential module $M \simeq D$ is projective and even free, thus torsion-free or 0-pure [24], being defined by the split exact sequence of free differential modules:

$$0 \to D \overset{\mathcal{D}_1}{\to} D^3 \overset{\mathcal{D}}{\to} D^3 \overset{\mathcal{D}_{-1}}{\to} D \to 0$$

We let the reader prove as an exercise that the adjoint sequence:

$$\begin{array}{c} 0 \leftarrow \theta \overset{ad(\mathcal{D}_{-1})}{\leftarrow} \nu \overset{ad(\mathcal{D})}{\leftarrow} \mu \overset{ad(\mathcal{D}_1)}{\leftarrow} \lambda \leftarrow 0 \\ 0 \leftarrow 1 \leftarrow 3 \leftarrow 3 \leftarrow 1 \leftarrow 0 \end{array}$$

starting from the Lagrange multiplier λ is also a split exact sequence of free differential modules.

We invite the reader to study, as a delicate exercise, the system of infinitesimal Lie equations defining the infinitesimal *unimodular* contact transformations preserving the 1-form $\alpha = dx^1 - x^3 dx^2$, thus also both the 2-form $\beta = d\alpha = dx^2 \wedge dx^3$ and the volume 3-form $\alpha \wedge \beta = dx^1 \wedge dx^2 \wedge dx^3$. Surprisingly, the Lie operator $\mathcal{D} : T \to \wedge^1 T^* \times_X \wedge^2 T^* : \xi \to (\mathcal{L}(\xi)\alpha = A, \mathcal{L}(\xi)\beta = B)$ for the geometric object $\omega = (\alpha, \beta)$ is involutive if and only if $d\alpha = c'\beta, d\beta = c''\alpha \wedge \beta$ with $c'c'' = 0$. It provides the differential Janet sequence $3 \to 6 \to 4 \to 1 \to 0$ with Euler-Poincaré characteristic $rk_D(M) = 3 - 6 + 4 - 1 = 0$. It follows that $\mathcal{D}$ cannot be parametrized. We have $\mathcal{D}_1 : (A, B) \to \left(dA - c'B = U, dB - c''(A \wedge \beta + \alpha \wedge B) = V\right)$ and $\mathcal{D}_2 : (U, V) \to dU + c'V = W$ because $c'c'' = 0$. (See [12] and the recent [19] for more details).

4. Einstein equations

If $g_1 \subset T^* \otimes T$ with $dim(g_1) = n(n-1)/2$ is the symbol of the Killing system $R_1 \subset J_1(T)$ with $dim(R_1) = n(n+1)/2$, its first prolongation is $g_2 = 0$. We may introduce the Riemann tensor $\rho = \left(\rho^k_{l,ij}\right) \in F_1 = H^2_1(g_1)$ with $n^2(n^2-1)/12$ components in the short exact sequence [13, 18, 25, 27–29]:

$$0 \to F_1 \to \wedge^2 T^* \otimes g_1 \overset{\delta}{\to} \wedge^3 T^* \otimes T \to 0 \tag{17}$$

Linearizing the *Ricci* tensor $\left(\rho_{ij} = \rho^r_{i,rj} = \rho_{ji}\right) \in S_2 T^*$ over the Minkowski metric ω, we obtain the usual second order homogeneous *Ricci* operator $\Omega \to R$ with 4 terms (*care*):

$$\begin{array}{c} 2R_{ij} = \omega^{rs}\left(d_{rs}\Omega_{ij} + d_{ij}\Omega_{rs} - d_{ri}\Omega_{sj} - d_{sj}\Omega_{ri}\right) = 2R_{ji} \\ tr(\Omega) = \omega^{ij}\Omega_{ij} \Rightarrow tr(R) = \omega^{ij}R_{ij} = \omega^{ij}d_{ij}tr(\Omega) - \omega^{ru}\omega^{sv}d_{rs}\Omega_{uv} \end{array}$$

We may define the *Einstein* operator by setting $\varepsilon_{ij} = \rho_{ij} - \frac{1}{2}\omega_{ij}\omega^{rs}\rho_{rs} \Rightarrow E_{ij} = R_{ij} - \frac{1}{2}\omega_{ij}tr(R)$ and obtain the 6 terms (*care*) [9]:

$$2E_{ij} = \omega^{rs}\left(d_{rs}\Omega_{ij} + d_{ij}\Omega_{rs} - d_{ri}\Omega_{sj} - d_{sj}\Omega_{ri}\right) - \omega_{ij}(\omega^{rs}\omega^{uv}d_{rs}\Omega_{uv} - \omega^{ru}\omega^{sv}d_{rs}\Omega_{uv})$$

We have the (locally exact) differential sequence of operators acting on sections of vector bundles:

$$\begin{array}{c} T \underset{1}{\overset{Killing}{\longrightarrow}} S_2T^* \underset{2}{\overset{Riemann}{\longrightarrow}} F_1 \underset{1}{\overset{Bianchi}{\longrightarrow}} F_2 \\ n \overset{\mathcal{D}}{\rightarrow} n(n+1)/2 \overset{\mathcal{D}_1}{\rightarrow} n^2(n^2-1)/12 \overset{\mathcal{D}_2}{\rightarrow} n^2(n^2-1)(n-2)/24 \end{array} \tag{18}$$

where $F_2 = H_1^3(g_1)$ is similarly defined by the short exact sequence:

$$0 \rightarrow F_2 \rightarrow \wedge^3 T^* \otimes g_1 \overset{\delta}{\rightarrow} \wedge^4 T^* \otimes T \rightarrow 0 \tag{19}$$

Our purpose is now to study the differential sequence onto which its right part is projecting:

$$\begin{array}{c} S_2T^* \underset{2}{\overset{Einstein}{\longrightarrow}} S_2T^* \underset{1}{\overset{div}{\longrightarrow}} T^* \rightarrow 0 \\ n(n+1)/2 \rightarrow n(n+1)/2 \rightarrow n \rightarrow 0 \end{array}$$

and the following adjoint sequence where we have set [13, 18, 25, 27–29]:

$$ad(T) \overset{Cauchy}{\longleftarrow} ad(S_2T^*) \overset{Beltrami}{\longleftarrow} ad(F_1) \overset{Lanczos}{\longleftarrow} ad(F_2) \tag{20}$$

In this sequence, if E is a vector bundle over the ground manifold X with dimension n, we may introduce the new vector bundle $ad(E) = \wedge^n T^* \otimes E^*$, where E^* is obtained from E by inverting the transition rules exactly like T^* is obtained from T. We have for example $ad(T) = \wedge^n T^* \otimes T^* \simeq \wedge^n T^* \otimes T \simeq \wedge^{n-1} T^*$ because T^* is isomorphic to T by using the metric ω. The 10×10 *Einstein* operator matrix is induced from the 10×20 *Riemann* operator matrix and the 10×4 *div* operator matrix is induced from the 20×20 *Bianchi* operator matrix. We advise the reader not familiar with the formal theory of systems or operators to follow the computation in dimension $n = 2$ with the 1×3 *Airy* operator matrix, which is the formal adjoint of the 3×1 *Riemann* operator matrix, and $n = 3$ with the 6×6 *Beltrami* operator matrix which is the formal adjoint of the 6×6 *Riemann* operator matrix which is easily seen to be self-adjoint up to a change of basis.

- $n = 2$: The stress equations become $d_1\sigma^{11} + d_2\sigma^{12} = 0, d_1\sigma^{21} + d_2\sigma^{22} = 0$. Their second order parametrization $\sigma^{11} = d_{22}\phi, \sigma^{12} = \sigma^{21} = -d_{12}\phi, \sigma^{22} = d_{11}\phi$ has been provided by George Biddell Airy in 1863 [36] and is well known [4]. We get the second order system:

$$\left\{\begin{array}{ll} \sigma^{11} & \equiv d_{22}\phi = 0 \\ -\sigma^{12} & \equiv d_{12}\phi = 0 \\ \sigma^{22} & \equiv d_{11}\phi = 0 \end{array}\right. \quad \begin{array}{|cc|} 1 & 2 \\ 1 & \bullet \\ 1 & \bullet \\ \hline \end{array}$$

which is involutive with one equation of class 2, 2 equations of class 1 and it is easy to check that the 2 corresponding first order CC are just the *Cauchy* equations. Of course, the *Airy* function (1 term) has absolutely nothing to do with the perturbation of the metric (3 terms). With more details, when ω is the Euclidean metric, we may consider the only component:

$$tr(R) = (d_{11} + d_{22})(\Omega_{11} + \Omega_{22}) - (d_{11}\Omega_{11} + 2d_{12}\Omega_{12} + d_{22}\Omega_{22}) = d_{22}\Omega_{11} + d_{11}\Omega_{22} - 2d_{12}\Omega_{12}$$

Multiplying by the Airy function ϕ and integrating by parts, we get $Airy = ad(Riemann)$ and $Cauchy = ad(Killing)$ in the following differential sequences:

$$2 \underset{1}{\overset{Killing}{\longrightarrow}} 3 \underset{2}{\overset{Riemann}{\longrightarrow}} 1 \to 0$$

$$0 \leftarrow 2 \underset{1}{\overset{Cauchy}{\longleftarrow}} 3 \underset{2}{\overset{Airy}{\longleftarrow}} 1$$

- $n = 3$: It is quite more delicate to parametrize the 3 PD equations:

$$d_1\sigma^{11} + d_2\sigma^{12} + d_3\sigma^{13} = 0, \quad d_1\sigma^{21} + d_2\sigma^{22} + d_3\sigma^{23} = 0, \quad d_1\sigma^{31} + d_2\sigma^{32} + d_3\sigma^{33} = 0$$

A direct computational approach has been provided by Eugenio Beltrami in 1892 [37, 38], James Clerk Maxwell in 1870 [39] and Giacinto Morera in 1892 [38, 40] by introducing the 6 *stress functions* $\phi_{ij} = \phi_{ji}$ in the *Beltrami parametrization*. The corresponding system:

$$\left\{\begin{array}{rl} \sigma^{11} \equiv & d_{33}\phi_{22} + d_{22}\phi_{33} - 2d_{23}\phi_{23} = 0 \\ -\sigma^{12} \equiv & d_{33}\phi_{12} + d_{12}\phi_{33} - d_{13}\phi_{23} - d_{23}\phi_{13} = 0 \\ \sigma^{22} = & d_{33}\phi_{11} + d_{11}\phi_{33} - 2d_{13}\phi_{13} = 0 \\ \sigma^{13} \equiv & d_{23}\phi_{12} + d_{12}\phi_{23} - d_{22}\phi_{13} - d_{13}\phi_{22} = 0 \\ -\sigma^{23} \equiv & d_{23}\phi_{11} + d_{11}\phi_{23} - d_{12}\phi_{13} - d_{13}\phi_{12} = 0 \\ \sigma^{33} \equiv & d_{22}\phi_{11} + d_{11}\phi_{22} - 2d_{12}\phi_{12} = 0 \end{array}\right. \quad \begin{array}{|ccc|} 1 & 2 & 3 \\ 1 & 2 & 3 \\ 1 & 2 & 3 \\ 1 & 2 & \bullet \\ 1 & 2 & \bullet \\ 1 & 2 & \bullet \end{array}$$

is involutive with 3 equations of class 3, 3 equations of class 2 and no equation of class 1. The three characters are thus $\alpha_2^3 = 1 \times 6 - 3 = 3 < \alpha_2^2 = 2 \times 6 - 3 = 9 < \alpha_2^1 = 3 \times 6 - 0 = 18$ and we have $dim(g_2) = \alpha_2^1 + \alpha_2^2 + \alpha_2^3 = 18 + 9 + 3 = 30 = dim(S_2T^* \otimes S_2T^*) - dim(S_2T^*) = 6 \times 6 - 6$ [22]. The 3 CC are describing the stress equations which admit therefore a parametrization ... but without any geometric framework, in particular without any possibility to imagine that the above second order operator is *nothing else but* the *formal adjoint* of the *Riemann operator*, namely the (linearized) Riemann tensor with $n^2(n^2 - 1)/2 = 6$ independent components when $n = 3$ [12, 13].

Breaking the canonical form of the six equations which is associated with the Janet tabular, we may rewrite the Beltrami parametrization of the Cauchy stress equations as follows, after exchanging the third row with the fourth row, keeping the ordering $\{(11) < (12) < (13) < (22) < (23) < (33)\}$:

$$\begin{pmatrix} d_1 & d_2 & d_3 & 0 & 0 & 0 \\ 0 & d_1 & 0 & d_2 & d_3 & 0 \\ 0 & 0 & d_1 & 0 & d_2 & d_3 \end{pmatrix} \begin{pmatrix} 0 & 0 & 0 & d_{33} & -2d_{23} & d_{22} \\ 0 & -d_{33} & d_{23} & 0 & d_{13} & -d_{12} \\ 0 & d_{23} & -d_{22} & -d_{13} & d_{12} & 0 \\ d_{33} & 0 & -2d_{13} & 0 & 0 & d_{11} \\ -d_{23} & d_{13} & d_{12} & 0 & -d_{11} & 0 \\ d_{22} & -2d_{12} & 0 & d_{11} & 0 & 0 \end{pmatrix} \equiv 0$$

as an identity where 0 on the right denotes the zero operator. However, if Ω is a perturbation of the metric ω, the standard implicit summation used in continuum mechanics is, when $n = 3$:

$$\begin{aligned}\sigma^{ij}\Omega_{ij} &= \sigma^{11}\Omega_{11} + 2\sigma^{12}\Omega_{12} + 2\sigma^{13}\Omega_{13} + \sigma^{22}\Omega_{22} + 2\sigma^{23}\Omega_{23} + \sigma^{33}\Omega_{33} \\ &= \Omega_{22}d_{33}\phi_{11} + \Omega_{33}d_{22}\phi_{11} - 2\Omega_{23}d_{23}\phi_{11} + \ldots \\ &\quad +\Omega_{23}d_{13}\phi_{12} + \Omega_{13}d_{23}\phi_{12} - \Omega_{12}d_{33}\phi_{12} - \Omega_{33}d_{12}\phi_{12} + \ldots\end{aligned}$$

because *the stress tensor density* σ *is supposed to be symmetric*. Integrating by parts in order to construct the adjoint operator, we get:

$$\begin{aligned}\phi_{11} &\rightarrow d_{33}\Omega_{22} + d_{22}\Omega_{33} - 2d_{23}\Omega_{23} \\ \phi_{12} &\rightarrow d_{13}\Omega_{23} + d_{23}\Omega_{13} - d_{33}\Omega_{12} - d_{12}\Omega_{33}\end{aligned}$$

and so on. The identifications $Beltrami = ad(Riemann), Lanczos = ad(Bianchi)$ in the diagram:

$$\begin{array}{ccccccccc} & 3 & \underset{1}{\overset{Killing}{\longrightarrow}} & 6 & \underset{2}{\overset{Riemann}{\longrightarrow}} & 6 & \underset{1}{\overset{Bianchi}{\longrightarrow}} & 3 & \rightarrow 0 \\ 0 \leftarrow & 3 & \underset{1}{\overset{Cauchy}{\longleftarrow}} & 6 & \underset{2}{\overset{Beltrami}{\longleftarrow}} & 6 & \underset{1}{\overset{Lanczos}{\longleftarrow}} & 3 & \end{array} \quad (21)$$

prove that the *Cauchy* operator has nothing to do with the *Bianchi* operator [12, 13].

When ω is the Euclidean metric, the link between the two sequences is established by means of the elastic constitutive relations $2\sigma_{ij} = \lambda tr(\Omega)\omega_{ij} + 2\mu\Omega_{ij}$ with the Lamé elastic constants (λ, μ) but mechanicians are usually setting $\Omega_{ij} = 2\varepsilon_{ij}$. Using the standard Helmholtz decomposition $\vec{\xi} = \vec{\nabla}\varphi + \vec{\nabla}\wedge\vec{\psi}$ and substituting in the dynamical equation $d_i\sigma^{ij} = \rho d^2/dt^2\xi^j$ where ρ is the mass per unit volume, we get the longitudinal and transverse wave equations, namely $\Delta\varphi - \frac{\rho}{\lambda+2\mu}\frac{d^2}{dt^2}\varphi = 0$ and $\Delta\vec{\psi} - \frac{\rho}{\mu}\frac{d^2}{dt^2}\vec{\psi} = 0$, responsible for earthquakes, with respective speeds $c_L^2 = (\lambda + 2\mu)/\rho$ and $c_T^2 = \mu/\rho$.

Then, taking into account, the factor 2 involved by multiplying the second, third and fifth row by 2, we get the new 6×6 operator matrix with rank 3 which is clearly self-adjoint:

$$\begin{pmatrix} 0 & 0 & 0 & d_{33} & -2d_{23} & d_{22} \\ 0 & -2d_{33} & 2d_{23} & 0 & 2d_{13} & -2d_{12} \\ 0 & 2d_{23} & -2d_{22} & -2d_{13} & 2d_{12} & 0 \\ d_{33} & 0 & -2d_{13} & 0 & 0 & d_{11} \\ -2d_{23} & 2d_{13} & 2d_{12} & 0 & -2d_{11} & 0 \\ d_{22} & -2d_{12} & 0 & d_{11} & 0 & 0 \end{pmatrix}$$

Following Maxwell, we keep $\phi_{11} = A, \phi_{22} = B, \phi_{33} = C$, set $\phi_{12} = \phi_{23} = \phi_{31} = 0$ and get:

$$\left\{\begin{array}{ll} \sigma^{11} \equiv & d_{33}B + d_{22}C = 0 \\ \sigma^{22} \equiv & d_{33}A + d_{11}C = 0 \\ -\sigma^{23} \equiv & d_{23}A = 0 \\ \sigma^{33} \equiv & d_{22}A + d_{11}B = 0 \\ -\sigma^{13} \equiv & d_{13}B = 0 \\ -\sigma^{12} \equiv & d_{12}C = 0 \end{array}\right. \quad \begin{array}{|ccc|} 1 & 2 & 3 \\ 1 & 2 & 3 \\ 1 & 2 & \bullet \\ 1 & 2 & \bullet \\ 1 & \bullet & \bullet \\ 1 & \bullet & \bullet \\ \hline \end{array}$$

This system may not be involutive and no CC can be found *"a priori"* because the coordinate system is surely not δ-regular. Effecting the linear change of coordinates $\bar{x}^1 = x^1, \bar{x}^2 = x^2, \bar{x}^3 = x^3 + x^2 + x^1$ and taking out the bar for simplicity, we obtain the involutive system:

$$\left\{\begin{array}{l} d_{33}C + d_{13}C + d_{23}C + d_{12}C = 0 \\ d_{33}B + d_{13}B = 0 \\ d_{33}A + d_{23}A = 0 \\ d_{23}C + d_{22}C - d_{13}C - d_{13}B - d_{12}C = 0 \\ d_{23}A - d_{22}C + d_{13}B + 2d_{12}C - d_{11}C = 0 \\ d_{22}A + d_{22}C - 2d_{12}C + d_{11}C + d_{11}B = 0 \end{array}\right. \quad \begin{array}{|ccc|} 1 & 2 & 3 \\ 1 & 2 & 3 \\ 1 & 2 & 3 \\ 1 & 2 & \bullet \\ 1 & 2 & \bullet \\ 1 & 2 & \bullet \\ \hline \end{array}$$

and it is easy to check that the 3 CC obtained just amount to the desired 3 stress equations when coming back to the original system of coordinates. However, the three characters are different as we have now $\alpha_2^3 = 3 - 3 = 0 < \alpha_2^2 = 2 \times 3 - 3 = 3 < \alpha_2^1 = 3 \times 3 - 0 = 9$ with sum equal to $dim\left(g_2\right) = 6 \times 3 - 6 = 18 - 6 = 12$. *We have thus a minimum parametrization*. Of course, *if there is a geometrical background, this change of local coordinates is hiding it totally*. Moreover, we notice that the stress functions kept in the procedure are just the ones on which d_{33} is acting. The reason for such an apparently technical choice is related to very general deep arguments in the theory of differential modules that will only be explained at the end of the paper.

Following Morera, we keep $\phi_{23} = L, \phi_{13} = M, \phi_{12} = N$, set $\phi_{11} = \phi_{22} = \phi_{33} = 0$, and get:

$$\left\{\begin{array}{l} d_{23}L = 0 \\ d_{33}N - d_{13}L - d_{23}M = 0 \\ d_{13}M = 0 \\ d_{22}M - d_{23}N - d_{12}L = 0 \\ d_{11}L - d_{12}M - d_{13}N = 0 \\ d_{12}N = 0 \end{array}\right.$$

Using the same change of coordinates, we obtain the following involutive system:

$$\left\{\begin{array}{l} d_{33}N + d_{23}N + d_{13}N + d_{12}N = 0 \\ d_{33}M + d_{13}M = 0 \\ d_{33}L + d_{23}L = 0 \\ (d_{23}N + d_{23}M - d_{23}L) + (d_{13}N - d_{13}M + d_{13}L) + d_{12}N = 0 \\ 2d_{23}M + (d_{13}N - d_{13}M - d_{13}L) + d_{12}M - d_{11}L = 0 \\ d_{22}M + (d_{12}N - d_{12}M - d_{12}L) + d_{11}L = 0 \end{array}\right. \quad \begin{array}{|ccc|} 1 & 2 & 3 \\ 1 & 2 & 3 \\ 1 & 2 & 3 \\ 1 & 2 & \bullet \\ 1 & 2 & \bullet \\ 1 & 2 & \bullet \\ \hline \end{array}$$

After elementary but tedious computations (*that could not be avoided*!), one can prove that the 3 CC corresponding to the 3 dots are effectively satisfied and that they correspond to the 3 Cauchy stress equations which are therefore parametrized. The parametrization is thus provided by an involutive operator defining a torsion module because the character α_2^3 is vanishing in δ-regular coordinates, just like before for the Maxwell parametrization. *We have thus another minimum parametrization*. Of course, such a result could not have been understood by Beltrami in 1892 because the work of Cartan could not be adapted easily in the language of exterior forms and the work of Janet appeared only in 1920 with no explicit reference to involution because only Janet bases are used while the Pommaret bases have only been introduced in 1978 (See [25, 29] for historical facts).

We may finally keeep $\{\phi_{11}, \phi_{12}, \phi_{22}\}$, set $\phi_{13} = \phi_{23} = \phi_{33} = 0$ and get the different involutive system with the same characters, *in particular* with $\alpha_2^3 = 0$:

$$\left\{\begin{array}{lcl} \sigma^{11} & \equiv & \partial_{33}\phi_{22} = 0 \\ -\sigma^{12} & \equiv & \partial_{33}\phi_{12} = 0 \\ \sigma^{22} & \equiv & \partial_{33}\phi_{11} = 0 \\ \sigma^{13} & \equiv & \partial_{23}\phi_{12} - \partial_{13}\phi_{22} = 0 \\ -\sigma^{23} & \equiv & \partial_{23}\phi_{11} - \partial_{13}\phi_{12} = 0 \\ \sigma^{33} & \equiv & \partial_{22}\phi_{11} + \partial_{11}\phi_{22} - 2\partial_{12}\phi_{12} = 0 \end{array}\right. \quad \begin{array}{|ccc|} 1 & 2 & 3 \\ 1 & 2 & 3 \\ 1 & 2 & 3 \\ 1 & 2 & \bullet \\ 1 & 2 & \bullet \\ 1 & 2 & \bullet \\ \hline \end{array} \tag{22}$$

So far, we have thus obtained three explicit local minimum parametrizations of the Cauchy stress equations with $n(n-1)/2 = 3$ stress potentials but there may be others.

- $n = 4$: It just remains to explain the relation of the previous results with Einstein equations. The first suprising link is provided by the following technical proposition:

Proposition 4.1: The Beltrami parametrization is just described by the *Einstein* operator when $n = 3$. The same confusion existing between the *Bianchi* operator and the *Cauchy* operator has been made by both Einstein and Beltrami because the *Einstein* operator is self-adjoint in arbitrary dimension $n \geq 3$, contrary to the *Ricci* operator.

Proof: The number of components of the Riemann tensor is $dim(F_1) = n^2(n^2-1)/12$. We have the combinatorial formula $n^2(n^2-1)/12 - n(n+1)/2 = n(n+1)(n+2)(n-3)/12$ expressing that the number of components of the Riemann tensor is always greater or equal to the number of components of the Ricci tensor whenever $n > 2$. Also, we have shown in many books [12, 13, 25, 27–29] or papers [16, 18] that the number of Bianchi identities is equal to $n^2(n^2-1)(n-2)/24$, that is 3 when $n = 3$ and 20 when $n = 4$. Of course, it is well known that the *div* operator, induced as CC of the *Einstein* operator, has n components in arbitrary dimension $n \geq 3$.

Accordingly, when $n = 3$ we have $n^2(n^2-1)/12 = n(n+1)/2 = 6$ and it only remains to prove that the *Einstein* operator reduces to the *Beltrami* operator *and not just to the Ricci operator*.

The following formulas can be found in any textbook on general relativity. In particular, the difference existing between R_{ij} (4 terms only) and E_{ij} (6 terms) can only be seen when $\omega_{i \neq j} = 0$. In our situation with $n = 3$ and the Euclidean metric, we obtain successively:

$$\begin{aligned}
2R_{12} = 2E_{12} &= (d_{11}+d_{22}+d_{33})\Omega_{12} + d_{12}(\Omega_{11}+\Omega_{22}+\Omega_{33}) \\
&\quad -(d_{11}\Omega_{12}+d_{12}\Omega_{22}+d_{13}\Omega_{23}) - (d_{12}\Omega_{11}+d_{22}\Omega_{12}+d_{23}\Omega_{13}) \\
&= d_{33}\Omega_{12}+d_{12}\Omega_{33}-d_{13}\Omega_{23}-d_{23}\Omega_{13} \\
2R_{11} &= (d_{11}+d_{22}+d_{33})\Omega_{11}+d_{11}(\Omega_{11}+\Omega_{22}+\Omega_{33}) \\
&\quad -2(d_{11}\Omega_{11}+d_{12}\Omega_{12}+d_{13}\Omega_{13}) \\
&= (d_{22}+d_{33})\Omega_{11}+d_{11}(\Omega_{22}+\Omega_{33})-2(d_{12}\Omega_{12}+d_{13}\Omega_{13}) \\
tr(R) &= (d_{11}\Omega_{22}+d_{11}\Omega_{33}+d_{22}\Omega_{11}+d_{22}\Omega_{33}+d_{33}\Omega_{11}+d_{33}\Omega_{22}) \\
&\quad -2(d_{12}\Omega_{12}+d_{13}\Omega_{13}+d_{23}\Omega_{23}) \\
2E_{11} &= d_{22}\Omega_{33}+d_{33}\Omega_{22}-2d_{23}\Omega_{23}
\end{aligned}$$

In the light of modern differential geometry, comparing these results with the works of Maxwell, Morera, Beltrami and Einstein, it becomes clear that they have been confusing the *div* operator induced from the *Bianchi* operator with the *Cauchy* operator. However, it is also clear that they both obtained a possibility to parametrize the *Cauchy* operator by means of 3 arbitrary potential like functions in the case of Maxwell and Morera, 6 in the case of Beltrami explaining the previous choices, and 10 in the case of Einstein. Of course, as they were ignoring that the *Einstein* operator was self-adjoint whenever $n \geq 3$, they did not notice that we have $Cauchy = ad(Killing)$ and they were unable to compare their results with the *Airy* operator found as early as in 1870 for the same mechanical purpose when $n = 2$. To speak in a rough way, the situation is similar to what could happen in the study of contact structures if one should confuse $\mathcal{D}_{-1}$ with $\mathcal{D}_1$ or $\mathcal{D}$ with $\mathcal{D}_2$ in the Killing sequence [16]. Finally, using the double differential duality test, we can choose a differential transcendence basis with $n(n-1)/2$ potentials that can be indexed by $\phi_{ij} = \phi_{ji}$ with $i<j$ or $1 \leq i,j \leq n-1$ or even $2 \leq i,j \leq n$ when the dimension $n \geq 2$ is arbitrary (See [24, 29] for more details on differential algebra).

□

Remark 4.2: The author of this paper is not an historian of sciences but a specialist of mathematical physics interested by the analogy existing between *electromagnetism* (EM), *elasticity* (EL), and *gravitation* (GR) by using the conformal group of space–time along the idea of H. Weyl [41] (See [4, 13, 18, 19, 28–31] for related works). It is thus difficult to imagine that Einstein could not have been aware of the works of Maxwell and Beltrami on the foundations of EL and tensor calculus as they were quite famous when he started his research on GR. We also notice that the Mach-Lippmann analogy [4, 13, 28, 29, 35] was introduced at the same time and that the phenomenological law of piezzo-electricity has been discovered by ... Maxwell [19, 20]. We do believe that classical variational calculus must be considered as a kind of "*duality theory*" that should only depend on new purely mathematical tools, namely "*group theory*" on one side and "*differential homological algebra*" on the other side (See [12, 13, 28] for the theory and [4] for the applications).

The two following crucial results, still neither known nor acknowledged today, are provided by the next proposition and corresponding corollary:

Proposition 4.3: The cauchy operator can be parametrized by the adjoint of the Ricci Operator

Proof: The *Einstein* operator $\Omega \to E$ is defined by setting $E_{ij} = R_{ij} - \frac{1}{2}\omega_{ij}tr(R)$ that we shall write $Einstein = C \circ Ricci$ where $C : S_2T^* \to S_2T^*$ is a symmetric matrix only depending on ω, which is invertible whenever $n \geq 3$. *Surprisingly*, we may also introduce the *same* linear transformation $C : \Omega \to \overline{\Omega} = \Omega - \frac{1}{2}\omega tr(\Omega)$ and the unknown composite operator $\mathcal{X} : \overline{\Omega} \to \Omega \to E$ in such a way that $Einstein = \mathcal{X} \circ C$ where $\mathcal{X}$ is defined by (See [15], 5.1.5, p. 134):

$$2E_{ij} = \omega^{rs} d_{rs}\overline{\Omega}_{ij} - \omega^{rs} d_{ri}\overline{\Omega}_{sj} - \omega^{rs} d_{sj}\overline{\Omega}_{ri} + \omega_{ij}\omega^{ru}\omega^{sv} d_{rs}\overline{\Omega}_{uv}$$

Now, introducing the test functions λ^{ij}, we get:

$$\lambda^{ij}E_{ij} = \lambda^{ij}\left(R_{ij} - \frac{1}{2}\omega_{ij}tr(R)\right) = \left(\lambda^{ij} - \frac{1}{2}\lambda^{rs}\omega_{rs}\omega^{ij}\right)R_{ij} = \overline{\lambda}^{ij}R_{ij}$$

Integrating by parts while setting as usual $\square = \omega^{rs} d_{rs}$, we obtain:

$$\left(\square\overline{\lambda}^{rs} + \omega^{rs} d_{ij}\overline{\lambda}^{ij} - \omega^{sj} d_{ij}\overline{\lambda}^{ri} - \omega^{ri} d_{ij}\overline{\lambda}^{sj}\right)\Omega_{rs} = \sigma^{rs}\Omega_{rs} \tag{23}$$

Moreover, suppressing the "bar" for simplicity, we have:

$$d_r\sigma^{rs} = \omega^{ij} d_{rij}\lambda^{rs} + \omega^{rs} d_{rij}\lambda^{ij} - \omega^{sj} d_{rij}\lambda^{ri} - \omega^{ri} d_{rij}\lambda^{sj} = 0 \tag{24}$$

As *Einstein* is a self-adjoint operator (contrary to the Ricci operator), we have the identities:

$$ad(Einstein) = ad(C) \circ ad(\mathcal{X}) \Rightarrow Einstein = C \circ ad(\mathcal{X}) \Rightarrow ad(\mathcal{X}) = Ricci \Rightarrow \mathcal{X} = ad(Ricci)$$

Indeed, $ad(C) = C$ because C is a symmetric matrix and we know that $ad(Einstein) = Einstein$. Accordingly, the operator $ad(Ricci)$ parametrizes the *Cauchy* equations, *without any reference* to the *Einstein* operator, which has no mathematical origin, in the sense that it cannot be obtained by any diagram chasing. The three terms after the *Dalembert* operator factorize through the divergence operator $d_i\lambda^{ri}$. We may thus add the *differential constraints* $d_i\lambda^{ri} = 0$ *without any reference to a gauge transformation* in order to obtain a (minimum) *relative parametrization* (see [24] for details and explicit examples). When $n = 4$ we finally obtain the adjoint sequences:

$$\begin{array}{ccccccc} & & 4 & \xrightarrow{Killing} & 10 & \xrightarrow{Ricci} & 10 \\ & & & & & & \\ 0 & \leftarrow & 4 & \xleftarrow{Cauchy} & 10 & \xleftarrow{ad(Ricci)} & 10 \end{array} \tag{25}$$

without any reference to the *Bianchi* operator and the induced *div* operator.

Finally, using Theorem 2.1 or Proposition 2.2, we may choose a differential transcendence basis made by $\{\lambda^{ij} | i < j\}$ or $\{\lambda^{ij} | 1 < i, j < n-1\}$ or even $\{\lambda^{ij} | 2 < i, j < n\}$ when $n \geq 2$.

□

According to **Theorem 3.2** and **Example 3.6**, the *Einstein* and thus also the *Ricci* operators cannot be parametrized. Now, according to **Corollary 3.3**, the differential module N that they define is *not* torsion-free.

Corollary 4.4: $t(N)$ is generated by the 10 components of the Weyl tensor and each component is killed by the Dalembert operator.

Proof: We first recall the 5 steps of the *double differential duality test* in this framework:

$\boxed{1} \Rightarrow$ Start with the *Einstein* operator $\mathcal{D}_1 : 10 \overset{Einstein}{\longrightarrow} 10$.

$\boxed{2} \Rightarrow$ Consider its formal adjoint: $ad(\mathcal{D}_1) : 10 \overset{Einstein}{\longleftarrow} 10$.

$\boxed{3} \Rightarrow$ Compute the generating CC, namely the Cauchy operator:
$ad(\mathcal{D}) : 4 \overset{Cauchy}{\longleftarrow} 10$.

$\boxed{4} \Rightarrow$ Consider its formal adjoint: $\mathcal{D} = ad(ad(\mathcal{D})) : 4 \overset{Killing}{\longrightarrow} 10$.

$\boxed{5} \Rightarrow$Compute the generating CC, namely the *Riemann* operator: $\mathcal{D}'_1 : 10 \overset{Riemann}{\longrightarrow} 20$.

With a slight abuse of language, we have the direct sum $Riemann = Ricci \oplus Weyl$ with $20 = 10 + 10$. It follows from differential homological algebra that the 10 additional CC in $\mathcal{D}'_1$ that are *not* in $\mathcal{D}_1$, are generating the torsion submodule $t(N)$ of the differential module N defined by the *Einstein* or *Ricci* operator. If K is a differential field and we consider the ring $D = K[d_1, \ldots, d_n] = K[d]$ of differential operators with coefficients in K, we know that $rk_D(\mathcal{D}) = rk_D(ad(\mathcal{D}))$ for any operator matrix $\mathcal{D}$ with coefficients in K. In the present situation, as the *Minkowski* metric has coefficients equal to $0, 1, -1$, we may choose $K = \mathbb{Q}$. Hence, there *must* exist operators $\mathcal{P}$ and $\mathcal{Q}$ with $rk_D(\mathcal{P}) = 10$ and:

$$\boxed{\mathcal{P} \circ Weyl = \mathcal{Q} \circ Ricci} \tag{26}$$

One may also notice that $rk_D(Einstein) = rk_D(Ricci)$ with:

$$rk_D(Einstein) = \frac{n(n+1)}{2} - n = \frac{n(n-1)}{2}, \quad rk_D(Riemann) = \frac{n(n+1)}{2} - n = \frac{n(n-1)}{2}$$

It is a pure chance that the differential ranks of the Einstein and Riemann operators are equal. Indeed, $rk_D(Einstein)$ has only to do with the *div* operator induced by contracting the *Bianchi* operator, while $rk_D(Riemann)$ has only to do with the classical *Killing* operator and the fact that the corresponding differential *Killing* module is a torsion module because we have a Lie group of transformations having $n + \frac{n(n-1)}{2} = \frac{n(n+1)}{2}$ parameters (translations + rotations). Hence, as the *Riemann* operator is a direct sum of the *Weyl* operator and the *Einstein* or *Ricci* operator according to the previous theorem, each component of the *Weyl* operator must be killed by a certain operator whenever the *Einstein* or *Ricci* equations in vacuum are satisfied. Equivalently, we have to prove that we obtain a torsion differential module if we set the 10 constraints $R_{ij} = 0$ in the 20 equations $R_{kl,ij} = 0$. With more details, we may start from the long exact sequence:

$$\boxed{0 \to \Theta \to 4 \underset{1}{\overset{Killing}{\longrightarrow}} 10 \underset{2}{\overset{Riemann}{\longrightarrow}} 20 \underset{1}{\overset{Bianchi}{\longrightarrow}} 20 \underset{1}{\longrightarrow} 6 \to 0} \tag{27}$$

This resolution of the set of *Killing* vector fields is *not at all* a *Janet* sequence because the *Killing* operator is not involutive as it is an operator of finite type with symbol of dimension $n(n-1)/2 = 6$ and one should need one prolongation for getting an involutive operator with vanishing symbol at order two. Splitting the *Riemann* operator, we get the commutative and exact diagram:

$$\begin{array}{ccccccccccc}
 & & & & 0 & & 0 & & 0 & & \\
 & & & & \downarrow & & \downarrow & & \downarrow & & \\
 & & 0 & & 10 & \longrightarrow & 16 & \longrightarrow & 6 & \longrightarrow & 0\\
 & & \downarrow & & \downarrow\uparrow & & \downarrow & & \| & & \\
4 & \overset{Killing}{\longrightarrow} & 10 & \overset{Riemann}{\longrightarrow} & 20 & \overset{Bianchi}{\longrightarrow} & 20 & \longrightarrow & 6 & \longrightarrow & 0\\
 & & \| & & \downarrow\uparrow & & \downarrow & & \downarrow & & \\
 & & 10 & \overset{Einstein}{\longrightarrow} & 10 & \overset{div}{\longrightarrow} & 4 & \longrightarrow & 0 & & \\
 & & \downarrow & & \downarrow & & \downarrow & & & & \\
 & & 0 & & 0 & & 0 & & & &
\end{array} \qquad (28)$$

Passing to the differential module point of view, we have the long exact sequence:

$$\boxed{0 \longrightarrow D^6 \longrightarrow D^{20} \overset{Bianchi}{\longrightarrow} D^{20} \overset{Riemann}{\longrightarrow} D^{10} \overset{Killing}{\longrightarrow} D^4 \longrightarrow M \longrightarrow 0} \qquad (29)$$

Which is a resolution of the *Killing* differential module $M = coker(Killing)$, and we have indeed the vanishing of the *Euler-Poincaré characteristic* with $rk_D(M) = 4 - 10 + 20 - 20 + 6 = 0$.

Accordingly, we have $N' = coker(Riemann) \simeq im(Killing) \subset D^4$ and thus N' is torsion-free with $rk_D(N') = 4 - 0 = 4 = n$ because $rk_D(M) = 0$. The kernel L of the epimorphism $N \longrightarrow N'$ is a torsion module because $rk_D(L) = rk_D(N) - rk_D(N') = 4 - 4 = 0$. As D is an integral domain and $N' \subset D^4$ is torsion-free, we have thus $L = t(N)$ in the following commutative and exact diagram where $N = coker(Einstein)$ is the differential module defined by Einstein equations in vacuum:

$$\begin{array}{ccccccccccc}
 & & & & & & & & & 0 & \\
 & & & & & & & & & \downarrow & \\
 & & 0 & & 0 & & 0 & & & t(N) & \\
 & & \downarrow & & \downarrow & & \downarrow & & & \downarrow & \\
 & 0 \longrightarrow & D^4 & \overset{div}{\longrightarrow} & D^{10} & \overset{Einstein}{\longrightarrow} & D^{10} & \longrightarrow & & N & \longrightarrow 0\\
 & \downarrow & \downarrow & & \downarrow & & \| & & & \downarrow & \\
0 \longrightarrow & D^6 \longrightarrow & D^{20} & \overset{Bianchi}{\longrightarrow} & D^{20} & \overset{Riemann}{\longrightarrow} & D^{10} & \longrightarrow & & N' & \longrightarrow 0\\
 & \| & \downarrow & & \downarrow & & \downarrow & & & \downarrow & \\
0 \longrightarrow & D^6 \longrightarrow & D^{16} & \longrightarrow & D^{10} & & 0 & & & 0 & \\
 & \downarrow & \downarrow & & \downarrow & & & & & & \\
 & 0 & 0 & & 0 & & & & & &
\end{array}$$

A snake chase in the previous diagram allows to exhibit the long exact connecting sequence:

$$\boxed{0 \longrightarrow D^6 \longrightarrow D^{16} \longrightarrow D^{10} \longrightarrow t(N) \longrightarrow 0}$$

We point out that, *for* $n = 4$ (*only*), the CC of the Weyl operator are of order 2 and *not* 1 like the Bianchi CC for the Riemann operator (See [12, 18], Appendix 2 for a computer algebra checking by A. Quadrat). Accordingly, we have the *conformal sequence* in which $\hat{\mathcal{D}}$ is the *conformal Killing operator* when $n = 4$, with $rk_D(\hat{M}) = 4 - 9 + 10 - 9 + 4 = 0$:

$$\boxed{0 \to \hat{\Theta} \to 4 \underset{1}{\overset{\hat{\mathcal{D}}}{\to}} 9 \underset{2}{\overset{\hat{\mathcal{D}}_1}{\to}} 10 \underset{2}{\overset{\hat{\mathcal{D}}_2}{\to}} 9 \underset{1}{\overset{\hat{\mathcal{D}}_3}{\to}} 4 \to 0} \qquad (30)$$

and *one cannot find canonical morphisms between the classical and the conformal resolutions* (!).

It follows from the last theorem that the short exact sequence $0 \to D^{10} \to D^{20} \to D^{10} \to 0$ splits with $D^{20} \simeq D^{10} \oplus D^{10}$ but the existence of a canonical lift $D^{20} \to D^{10} \to 0$ in the above diagram does not allow to split the right column and thus $N \neq N' \oplus t(N)$ as N' is not even free. As for the torsion elements, we have $t(N) = coker(D^{16} \to D^{10})$ and we may thus represent them by the 10 components of the Weyl tensor. *It is not at all evident* that $\Box$ is killing each component of the *Weyl* tensor whenever $R_{ij} = 0$. A tricky technical computation can be found in ([32], p. 206), ([33], Exercise 7.7) and ([13], p. 95).

Indeed, according to the double differential duality test, each additional CC in $\mathcal{D}'_1$ which is not already in $\mathcal{D}_1$ is a torsion element as it belongs to this module. Now, as $rk_D(\mathcal{D}) = rk_D(im(\mathcal{D})$ the differential ranks of the Einstein and Riemann operators are thus equal to $n(n-1)/2$, but *this is a pure coincidence* because $rk_D(Einstein)$ has only to do with the *div* operator induced by contracting the Bianchi identities, while $rk_D(Riemann)$ has only to do with the classical Killing operator and the fact that the corresponding differential module is a torsion module because we have a Lie group of transformations having $n + \frac{n(n-1)}{2} = \frac{n(n+1}{2}$ parameters (translations + rotations). Hence, as the Riemann operator is a direct sum of the Weyl operator and the Einstein or Ricci operator according to the previous theorem, each component of the Weyl operator must be killed by a certain operator whenever the Einstein or Ricci equations in vacuum are satisfied in arbitrary dimension $n \geq 4$.

We prove directly that each component of the Weyl tensor is killed by the Dalembertian.

With Christoffel symbols γ and the corresponding covariant derivative ∇ we know that $\nabla\omega = 0$ and we may thus move up and down the indices as needed. We provide this tricky computation using essentially the (second) Bianchi identities. We have successively:

$$\nabla_r \rho^k_{l,ij} + \nabla_i \rho^k_{l,jr} + \nabla_j \rho^k_{l,ri} = 0 \quad \Rightarrow \quad \nabla^r \rho_{rl,ij} - \nabla_i \rho_{lj} + \nabla_j \rho_{li} = 0$$
$$\Rightarrow \nabla^r \nabla_r \rho_{kl,ij} + \nabla^r \nabla_i \rho_{kl,jr} + \nabla^r \nabla_j \rho_{kl,ri} = 0$$
$$\Rightarrow \quad \nabla^r \nabla_r \rho_{kl,ij} + \nabla_i \nabla^r \rho_{kl,jr} + \nabla_j \nabla^r \rho_{kl,ri} + [\nabla^r, \nabla_i]\rho_{kl,jr} + [\nabla^r, \nabla_j]\rho_{kl,ri} = 0$$
$$\Rightarrow \quad \nabla^r \nabla_r \rho_{kl,ij} + \nabla_i \nabla^r \rho_{kl,jr} + \nabla_j \nabla^r \rho_{kl,ri} + \left(\sum quadratic\right) = 0$$

But we have also $\rho_{kl,ij} = \rho_{ij,kl}$ and thus
$\nabla^r \rho_{ij,rl} = \nabla_i \rho_{lj} - \nabla_j \rho_{li} \quad \Rightarrow \quad \nabla^r \rho_{jr} - \frac{1}{2}\nabla_j \rho = 0.$

$$\Box \rho_{kl,ij} = \left(\nabla_i\left(\nabla_k \rho_{lj} - \nabla_l \rho_{kj}\right)\right) - (i \leftrightarrow j) + \left(\sum quadratic\right)$$

Linearizing at the Euclidean metric for $n = 2, 3$ or at the Minkowski metric for $n = 4$, we get:

$$\Box R_{kl,ij} = d_i\left(d_k R_{lj} - d_l R_{kj}\right) - d_j\left(d_k R_{li} - d_l R_{ki}\right) \tag{31}$$

We may finally use the splitting formula for defining the Weyl tensor $\sigma^k_{l,ij}$ with $\sigma^r_{l,rj} = 0$, namely:

$$\sigma^k_{l,ij} = \rho^k_{l,ij} - \frac{1}{(n-2)}\left(\delta^k_i \rho_{lj} - \delta^k_j \rho_{li} + \omega^{ks}\left(\omega_{lj}\rho_{si} - \omega_{li}\rho_{sj}\right)\right) + \frac{1}{(n-1)(n-2)}\left(\delta^k_i \omega_{lj} - \delta^k_j \omega_{li}\right)\rho$$

$$\sigma^k_{l,ij} = \rho^k_{l,ij} - \left(\sum \rho_{rs}\right) \quad \Rightarrow \quad \Sigma^k_{l,ij} = R^k_{l,ij} - \left(\sum R_{rs}\right)$$

At any moment, we could have used the Ricci operator in place of the Einstein operator. □

Finally, for the sake of completeness, we compute *directly* the characters of the Einstein system.

Using a direct checking with the ordering 11 < 12 < 13 < 14 < 22 < ... < 34 < 44, we obtain:

$$E_{33} = \omega^{44} d_{44}\Omega_{33} + lower\ \ terms, \ E_{23} = \omega^{44} d_{44}\Omega_{23} + \ ...$$

We are in the position to compute the characters of the *Einstein* operator but a similar procedure could be followed with the *Ricci* operator. We obtain at once:

$$\beta^4_2 = 6, \beta^3_2 = 4, \beta^2_2 = 0, \beta^1_2 = 0 \Leftrightarrow \alpha^4_2 = 4, \alpha^3_2 = 16, \alpha^3_2 = 30, \alpha^1_2 = 40$$

a result leading to $dim(g_2) = \alpha^1_2 + \alpha^2_2 + \alpha^3_2 + \alpha^4_2 = 90$ and $dim(g_3) = \alpha^1_2 + 2\alpha^2_2 + 3\alpha^3_2 + 4\alpha^4_2 = 164$ in a coherent way with the long exact sequences:

$$0 \rightarrow g_2 \rightarrow S_2T^* \otimes S_2T^* \rightarrow S_2T^* \rightarrow 0, \ 0 \rightarrow g_3 \rightarrow S_3T^* \otimes S_2T^* \rightarrow T^* \otimes S_2T^* \rightarrow T^* \rightarrow 0$$

Now, we have by definition $div = (d_1, d_2, d_3, d_4)$ and $div \circ Einstein = (0,0,0,0)$. However, the *Einstein* operator is a 10×10 operator matrix which is self-adjoint up to a change of basis [2] because it is made with homogeneous second order terms. It is thus of rank 6 and we obtain therefore, with a slight abuse of language, $det(Einstein) = 0$. This result which is not evident at first sight must be compared with the Poincaré situation when $n = 3$:

$$(d_1 d_2 d_3)\begin{pmatrix} 0 & -d_3 & d_2 \\ d_3 & 0 & -d_1 \\ -d_2 & d_1 & 0 \end{pmatrix} = (0\,0\,0)$$

This module interpretation may thus question the proper origin and existence of gravitational waves because the *div* operator on the upper left part of the diagram has *strictly nothing to do* with the *Cauchy* $= ad(Killing)$ operator which cannot even appear *anywhere* in this diagram. Also, looking back to Example 3.3, if we use $ad(Einstein)$ or $ad(Ricci)$ in order to parametrize the Cauchy operator, the 10 potentials involved are similar to the Airy or Maxwell/Morera potentials and have thus *strictly nothing to do* with a perturbation of the metric. Such a result is explaining the *conceptual confusion* announced in the abstract.

Corollary 4.5: When $\mathcal{D}$ is a Lie operator of *finite type*, that is when $[\Theta, \Theta] \subset \Theta$ under the ordinary bracket of vector fields and $g_{q+r} = 0$ for r large enough, then the Spencer sequence is locally isomorphic to the tensor product of the Poincaré sequence for the exterior derivative by a finite dimensional Lie algebra $\mathcal{G}$, namely [29]:

$$0 \to \Theta \to \wedge^0 T^* \otimes \mathcal{G} \xrightarrow{d} \wedge^1 T^* \otimes \hat{A}\mathcal{G} \xrightarrow{d} \wedge^2 T^* \otimes \mathcal{G} \xrightarrow{d} \dots \xrightarrow{d} \wedge^n T^* \otimes \mathcal{G} \to 0. \quad (32)$$

It is thus formally exact both with its adjoint sequence. As it is known that the extension modules do not depend on the resolution used, this is the reason for which not only the Cauchy stress operator can be parametrized but also the *Cosserat couple-stress operator* $ad(D_1)$ can be parametrized by $ad(D_2)$, a result not evident at all (see [34, 35] for explicit computations). Similarly, in the case of the conformal Killing system $\hat{R}_1 \subset J_1(T)$ for $n = 4$, *the symbols do not depend on any conformal factor* because $\hat{g}_1$ is defined by $\omega_{rj}\xi_i^r + \omega_{ir}\xi_j^r - \frac{2}{n}\omega_{ij}\xi_r^r = 0$, $\hat{g}_2 \simeq T^*$ (the so-called *elations* of E. Cartan) is defined by $\xi_{ij}^k - \frac{1}{n}\left(\delta_i^k \xi_{rj}^r + \delta_j^k \xi_{ri}^r - \omega_{ij}\omega^{ks}\xi_{rs}^r\right) = 0$ (with the Kronecker notation) and finally $\hat{g}_3 = 0, \forall n \geq 3$. The EM field is thus a section $F \in \delta(T^* \otimes \hat{g}_2) = \delta(T^* \otimes T^*) = \wedge^2 T^*$ killed by the exterior derivative d. It follows that EM only depends on the conformal group and *not* on $U(1)$ [30, 31] in a coherent way with the dream of H. Weyl because $T^* \otimes T^* \simeq S_2T^* \oplus \wedge^2 T^* = (R_{ij}) \oplus (F_{ij})$ [29, 41].

Remark 4.6: A similar situation is well known for the *Cauchy- Riemann* equations when $n = 2$. Indeed, any infinitesimal complex transformation ξ must be solution of the linear first order homogeneous system $\xi_2^2 - \xi_1^1 = 0, \xi_2^1 + \xi_1^2 = 0$ of infinitesimal Lie equations which is defining a torsion differential module because we have $\xi_{11}^1 + \xi_{22}^1 = 0, \xi_{11}^2 + \xi_{22}^2 = 0$, that is ξ^1 and ξ^2 are *separately* killed by the *Laplace* operator $\Delta = d_{11} + d_{22}$.

Remark 4.7: A similar situation is also well known for the EM field F in electromagnetism. Indeed, starting with the first set of *Maxwell* equations $dF = 0$ (M_1) and using the *Minkowski constitutive law* in vacuum with electric constant ε_0 and magnetic constant μ_0, such that $\varepsilon_0\mu_0 c^2 = 1$ for the second set of *Maxwell* equations (M_2) for the induction $\mathcal{F}$, a standard tricky differential elimination allows to avoid the *Lorenz* (*no* "t") gauge condition for the EM potential. Indeed, from M_1 we get formally $d_r F_{ij} + d_i F_{jr} + d_j F_{ri} = 0$ and from M_2 we get $d_r \mathcal{F}^{ri} = 0 \Rightarrow d^r F_{ri} = 0$. We finally obtain *directly* $\Box F = d^r d_r F_{ij} = d^r(d_i F_{rj} - d_j F_{ri}) = 0$ [4].

Example 4.8: (*Lorenz condition for EM*)

We prove that the Lorenz gauge condition for EM is just amounting to a relative minimum parametrization. Using an Euclidean metric instead of the Minkowski metric in order to have the Minkowski constitutive relation $\mathcal{F} = F$ in vacuum between the EM field and the EM induction, we have the parametrization $d_i A_j - d_j A_i = F_{ij}$ and obtain the conservation of current through the composition:

$$d_i(d^i A^j - d^j A^i) = d_i d^i A^j - d^j(d_i A^i) = \mathcal{J}^j \Rightarrow d_j(d_i d^i A^j - d^j d_i A^i) = d_j \mathcal{J}^j = 0$$

with implicit summation on i and j. The differential module defined by the involutive system $d_{ii}A_j - d_{ij}A_i = 0$ has a differential rank equal to 1 as there is only one CC. Adding the Lorenz condition $d_i A^i = 0$ is bringing the rank to zero. The idea is just to prove that the inhomogeneous system $d_i d^i A^j = \mathcal{J}^j, d_i A^i = 0$ has *again* the only CC $d_j \mathcal{J}^j = 0$ like in **Example 1.3**.

5. Conclusion

After teaching elasticity theory during 25 years to students in some of the best french civil engineering schools, the author of this paper still keeps in mind one of the most fascinating but most difficult exercises that he has set up. The purpose was to explain why the dome of the cathedral St Paul in London is called "whispering cupola", that is why if you go up to the gallery at one point, you can listen to a friend whispering 30 meters away on the opposite side. This striking phenomena has been first studied by Lord Rayleigh in 1878 and he introduced in 1910 the "surface elastic waves" now called Rayleigh waves (See [3] p. 199). In fact the molecules close to the free surface are moving along small ellipses as a combination of standard longitudinal (L) and transverse (T) waves while propagating at a slightly smaller speed. Such waves cannot propagate in liquid but may travel all around the earth surface many times after earthquakes. If λ and μ are the elastic Lamé constants of the material with mass ρ per unit volume, the respective speeds are such that $c_L^2 = (\lambda + 2\mu)/\rho$, $c_T^2 = \mu/\rho$ and the speed of the Rayleigh wave is $c_R^2 = \chi(1 - 2\nu)/(2(1 - \nu))$ where $\nu = \lambda/(2(\lambda + \mu))$ is the Poisson coefficient. After (very) tedious computations one may find that χ must be the only real root of the cubic equation $\chi^3 - 8\chi^2 + 8\frac{2-\nu}{1-\nu}\chi - \frac{8}{1-\nu} = 0$ existing in the interval $[0, 1]$ because $0 < \nu < 0,5$.

Accordingly, elasticity can be considered as a way to parametrize the Cauchy operator when $n = 3$ while GR can be considered as a way to parametrize the Cauchy operator when $n = 4$. Hence, the situations existing with the Cauchy stress equations, with the Cosserat couple-stress equations and with the Maxwell equations are similar, only the constitutive laws are different. Meanwhile, we have shown why the mathematical foundations of conformal geometry must be revisited in this new framework which is valid in arbitrary dimension and could provide an intrinsic way to unify EM and GR along the dream of Weyl [12, 31].

The situation of the gravitational waves equations seems quite different but the least that can be said is that it is not coherent with *differential double duality*. However, it follows that *exactly the same confusion* has been done by Maxwell, Morera, Beltrami, and Einstein because, in all these cases, the operator considered is self-adjoint. Like the Michelson and Morley experiment, we do believe that Einstein already knew the previous works of all these researchers who were quite famous at the time he was active. In any case, the comparison of the various parametrizations described in this paper needs no comment.

Author details

Jean-Francois Pommaret
CERMICS, Ecole des Ponts ParisTech, France

*Address all correspondence to: jean-francois.pommaret@wanadoo.fr

References

[1] Pommaret J-F. Dualité Différentielle et Applications. Comptes Rendus Académie des Sciences Paris, Série I. 1995;**320**:1225-1230

[2] Pommaret J-F. The mathematical foundations of general relativity revisited. Journal of Modern Physics. 2013;**4**:223-239. DOI: 10.4236/jmp.2013.48A022

[3] Bellet D, Duc J. Problèmes d' Elasticité. Toulouse: Cepadues-Edition. 1977. ISBN 2.85428.018.0

[4] Pommaret J-F. Partial Differential Control Theory. Dordrecht: Kluwer; 2001 (Zbl 1079.93001)

[5] Kashiwara M. Algebraic study of systems of partial differential equations. mémoires de la société mathématique de france. 1995;**63** (Transl. from Japanese of his 1970 [master thesis]). p.1-72

[6] Goldschmidt H. Prolongations of linear partial differential equations: I inhomogeneous equations. Ann. Scient. Ec. Norm. Sup. 1968;**4**:617-625. DOI: 10.24033/asens.1173

[7] Spencer DC. Overdetermined systems of partial differential equations. Bulletin of the American Mathematical Society. 1965;**75**:1-114

[8] Hu S-T. Introduction to Homological Algebra. San Francisco: Holden-Day; 1968

[9] Northcott DG. An Introduction to Homological Algebra. Cambridge: CambridgeUniversity Press; 1966

[10] Rotman JJ. An Introduction to Homological Algebra. Urbana: Academic Press; 1979

[11] Schneiders J-P. An introduction to D-modules. Bulletin de la Société Royale des Sciences de Liège. 1994;**63**:223-295

[12] Pommaret J-F. Deformation Theory of Algebraic and Geometric Structures. Saarbrucken, Germany: Lambert Academic Publisher (LAP); 2016. http://arxiv.org/abs/1207.1964

[13] Pommaret J-F. New Mathematical Methods for Physics. Mathematical Physics Books: Nova Science Publishers, New York; 2018. p. 150

[14] Bieri L, Garfinkle D, Hill CD, Nurowski P, Yunes N. The Mathematics of Gravitational Waves. Notices of the AMS. 2017;**64**(7):686-707

[15] Foster J, Nightingale JD. A Short Course in General Relativity. New York: Longman; 1979

[16] Pommaret J-F. Killing operator for the Kerr metric. Available from: https://arxiv.org/abs/2211.00064)

[17] Lanczos C. The splitting of the Riemann tensor. Reviews of Modern Physics. 1962;**34**:379-389

[18] Pommaret J-F. Homological solution of the lanczos problems in arbitrary dimension. Journal of Modern Physics. 2021;**12**:829-858. DOI: 10.4236/jmp.2020.1110104

[19] Pommaret J-F. How many structure constants do exist in Riemannian geometry. Mathematics in Computer Science. 2022;**16**:23. DOI: 10.1007/s11786-022-00546-3

[20] Pommaret J-F. Algebraic analysis of control systems defined by partial differential equations. In: "Advanced

Topics in Control Systems Theory", Lecture Notes in Control and Information Sciences. Vol. 311. London: Springer; 2005. Chapter 5. pp. 155-223

[21] Pommaret J-F, Quadrat A. Localization and parametrization of linear multidimensional control systems. Systems & Control Letters. 1999;**37**: 247-260

[22] Pommaret J-F. Why gravitational waves cannot exist. Journal of Modern Physics. 2017;8: 2122-2158. Available from: https://arxiv.org/abs/1708.06575. DOI: 104236/jmp.2017.813130

[23] Pommaret J-F. Minimum parametrization of the cauchy stress operator. Journal of Modern Physics. 2021;12:453-482. Available from: https://arxiv.org/abs/2101.03959. DOI: 10.4236/jmp.2021.124032

[24] Pommaret J-F. Relative parametrization of linear multidimensional systems. Multidimensional Systems and Signal Processing. 2015;**26**:405-437. DOI: 10.1007/s11045-013-0265-0

[25] Pommaret J.-F.: Systems of Partial Differential Equations and Lie Pseudogroups, Gordon and Breach, New York (1978); Russian translation: MIR, Moscow. 1983

[26] Zerz E. Topics in multidimensional linear systems theory. In: Lecture Notes in Control and Information Sciences, LNCIS. Vol. 256. London: Springer; 2000

[27] Pommaret J-F. Differential Galois Theory. New York: Gordon and Breach; 1983

[28] Pommaret J-F. Lie Pseudogroups and Mechanics. New York: Gordon and Breach; 1988

[29] Pommaret J-F. Partial Differential Equations and Group Theory. Kluwer; 1994. DOI: 10.1007/978-94-017-2539-2

[30] Pommaret J-F. Spencer operator and applications: From continuum mechanics to mathematical physics. In: Continuum Mechanics-Progress in Fundamentals and Engineering Applications. Gan Y, editor. London: InTech; 2012. Available from: DOI: 10.5772/35607. ISBN: 978-953-51-0447–6

[31] Pommaret J-F. Nonlinear conformal electromagnetism. Available from: https://arxiv.org/abs/2007.01710. DOI: 10.4236/jmp.2022.134031

[32] Choquet-Bruhat Y. Introduction to General Relativity, Black Holes and Cosmology. Oxford: Oxford University Press; 2015

[33] Hughston LP, Tod KP. An Introduction to General Relativity, London Math. Soc. Students Texts 5. Cambridge: Cambridge University Press; 1990

[34] Cosserat E, Cosserat F. Théorie des Corps Déformables. Paris: Hermann; 1909

[35] Pommaret J-F. Parametrization of Cosserat Equations. Acta Mechanica. 2010;**215**:43-55. DOI: 10.1007/s00707-010-0292-y

[36] Airy GB. On the strains in the interior of beams. Philosophical Transactions. Royal Society of London. 1863;**153**:49-80

[37] Beltrami E. Osservazioni sulla Nota Precedente. Atti della Accademia Nazionale dei Lincei Rend. 1892;**1**(5): 137-141 + 233-234 Collected Works, t IV

[38] Landriani GS. A note about an exchange of opinions between Beltrami and Morera on elastic equilibrium equations. Meccanica. 2017;**52**: 2801-2806. DOI: 10.1007/s11012-016-0611-z

[39] Maxwell JC. On reciprocal figures, frames and diagrams of forces. Transactions of the Royal Society of Edinburgh. 1870;**26**:1-40

[40] Morera G. Soluzione Generale della Equazioni Indefinite di Equilibrio di un Corpo Continuo. Atti della Academia Nazionale dei Lincei Rend. 1892;**1**(5): 137-141 + 233-234

[41] Weyl H. Space, Time, Matter, (1918). New York: Dover; 1952

Chapter 2

Gravitational Waves, Fields, and Particles in the Frame of (1 + 4)D Extended Space Model

Dmitry Yu Tsipenyuk and Wladimir B. Belayev

Abstract

Interval s is used as a fifth additional coordinate. We employ (1 + 4)-dimensional space G, an extension of the (1 + 3)-dimensional Minkowski space M. Interval changes under the transformations in the extended space G. From a physical point of view our expansion means that processes in which the rest mass of the particle's changes are acceptable now. In the Extended space model (ESM), a photon can have a nonzero variable mass. The equations for the plane-wave potentials describing the process of its localization are presented and their solution is found.

Keywords: (1 + 4)D space model, wave-particle interaction, plane wave localization, Lagrange mechanics, photon gravitational mass, graviton

1. Introduction

Since the late nineteenth century, there have been discussions about the issue of integrating electromagnetic and gravitational forces into a single field. These attempts have been made by constructing geometric models of physical interactions and interpretation of physics as geometry in the spaces of a larger number of dimensions. F. Klein [1] developed the Hamilton-Jacobi theory in the late nineteenth century as optics in the space with more dimensions. His thoughts did not, however, evolve at that time. By developing the general theory of relativity (GRT), interest in the issue of geometrization of physics has recently increased [2]. There have been attempts to use gravity as an analogy to define electromagnetic in geometric terms.

Instead of attempting to develop a new model, their authors tried to improve the GRT approach that has already existed. The most well known were the T. Kaluza [3] and O. Klein [4] models. The works of H. Mandel [5] and V. Fock [6] are also remarkable. The fact that they could only use a five-dimensional space is remarkable. The issue of the fifth coordinate's physical interpretation has not yet been satisfactorily resolved. The development of these approaches was attempted by scientists and researchers, including Einstein [7], de Broglie, Gamow, and Rumer [8]; nevertheless, they were unsuccessful in generating any interesting results. We believe that the reason is because, without including novel physical concepts, their works were formal expansions of previous models. Another area of the geometrization of physical interactions that we should highlight is the theory of gauge fields [9]. According to this paradigm, all interactions: electromagnetism, gravity, and others are viewed geometrically [10].

Afterward, in arrangement to form the hypothesis of basic particles, another approach to combining gravity with other interactions was developed. K.P. Stanyukovich and M.A. Markov suggested attempts to account for the gravitational field in the description of the interaction of elementary particles 50 years ago [11, 12]. They proposed the idea of two types of heave particles-planckions and maximons.

Three essential constants in nature were supposed by the authors: the Planck constant $\hbar$, lights speed c, gravitational constant G. Above values can be used to build expressions with dimensions of time, length, and mass. They are called Planck time t_{Pl}, Planck length l_{Pl}, and Planck mass m_{Pl}:

$$t_{Pl} = \sqrt{\frac{\hbar G}{c^5}} \sim 10^{-43} sec\, , l_{Pl} = \sqrt{\frac{\hbar G}{c^3}} \sim 10^{-33} cm, m_{Pl} = \sqrt{\frac{\hbar c}{G}} \sim 10^{-5} g \tag{1}$$

A particle with mass m corresponds to the Compton wavelength in the quantum theory

$$\lambda_c = \frac{\hbar}{mc}. \tag{2}$$

Particle size can be associated with this wavelength. In the case of the Planck mass m_{Pl} substituted in the formula (2), it turns out that the Compton wavelength coincides with the Planck length l_{Pl}.

$$\lambda_c = l_{Pl}. \tag{3}$$

But another linear parameter can be associated with mass m—the Schwarzschild gravitational radius

$$r_g = \frac{Gm}{c^2} \tag{4}$$

A spherically symmetric distribution of matter, according to the GRT, collapses into a black hole when it is squeezed to such a size. As a result, it is presently assumed that is the maximum value of an elementary particle's mass m_{Pl}. Such particles were named maximons. Large mass particles might become black holes. The corresponding gravitational radius r_{gr} can be considered as the minimum elementary particles' possible size. If we substitute the Planck mass m_{Pl} in the formula (4), we will take the result

$$r_{gr} = 2\sqrt{\frac{G}{c^3}} = 2l_{Pl} \tag{5}$$

Thus, the gravitational radius of maximon coincides in order of magnitude with Planck length. In Landau's work [13], estimates for the value of the "radius" of elementary particles were obtained, based on the limit of applicability of electrodynamics representations in quantum mechanics. Interestingly, the "radius" of the electron at the same time was equal to zero. Such relations were discussed in an attempt to take into account the gravitational forces in the processes of interaction of elementary particles. This approach assumes the initial existence of particles with a large rest mass, and since we do not observe such objects, it is not clear how it can be used to describe the processes occurring in the laboratory.

The name extended space model (ESM) is due to the fact that it is formulated in a flat five-dimensional space G(1,4) with the metric (+− − − −). ESM approach [14–17] is fundamentally different from all these and similar theories. ESM is based on the physical hypothesis that mass (rest mass) and its conjugation (interval) are dynamic variables, and its values are determined by the field-particle interactions. So, ESM is a direct SRT generalization. Interval and rest mass are invariants but can be changed in ESM. In particular, photon mass can be positive and negative. This mass can vary because of electromagnetic interactions and generates gravity interactions. This situation allows us to consider gravity and electromagnetism as unit fields. The equations for the plane-wave potentials describing the process of its localization are presented. The exact solution of these equations is found.

In GRT, the definition of the momenta of material and light particles moving in curvilinear space-time, and the forces acting on them, aims to find relativistic corrections to Newton's theory of gravitation for a weak gravitational field. In [18–20], the second derivatives of the coordinates along the path are considered as components of the 4-vector of the force acting on a material particle of a unit mass. However, in the gravitational field, not only the 4-momentum of matter alone, but the 4-momentum of matter together with the gravitational field should be preserved [18]. In the equation of particle motion containing force in this form, there is no energy and momentum transferred to the gravitational field.

Another approach is the choice of the Lagrangian of the particle, the definition of generalized forces as its partial derivatives with respect to the coordinate in accordance with Lagrange mechanics [21–24]. In GRT, the physical velocities of particles are associated with the components of the contravariant 4-velocity vector. Therefore, the physical force is aligned with the upper index vector associated with the generalized force vector. The energy and momenta of particles are considered to be the components of the contravariant 4-vector of energy-momentum, as is done in [18] for a particle moving in the Minkovsky space-time. The equations of motion will contain an additional term, which express the rate of change of the energy and momentum acquired by the gravitational field when a particle moves in it.

In the Fock proof [25] of the light motion along geodesics, the time component of the covariant 4-velocity vector is taken as the Hamiltonian. Application of the variational principle of the energy stationary integral to the motion of a light-like [21–24] particle in a gravitational field does not lead to a violation of the isotropy of the light path. In the generalized Fermat's principle [10], a variation of the integral of the time component of the 4-velocity vector is used and gives the trajectory of light movement that coincides with the geodesic.

For weak gravity, the analogy of the particle dynamics in Schwarzschild space-time with Newtonian gravitational theory permits to determine the passive gravitational mass of a photon. It is equal to twice the material particle mass of the same energy corresponded to non-gravitational interactions. This agrees with the results of Tolman, Ehrenfest, and Podolsky for the photon effective gravitational mass in the interaction between light packets or beams and matter particles [26, 27]. Observance of conservation of energy as a gravitation source suggests that at annihilation of an electron and positron in addition to gamma quanta, the particles g^{-} are released [22–24]. The birth of gamma-ray electron-positron pair leads to the appearance of a particle, which is identified [28] as a graviton.

2. ESM short description

According to the ESM, several physical values that are assumed to be constant under the conventional method are really not constant and can now change depending on the circumstances. We are discussing the zero mass of the photon as well as the rest mass of heavy particles. A model similar to the ESM model was developed by Wesson and his coauthors [29–31]. Wesson proposed to use "mass" as the fifth coordinate, in addition to the time and three spatial coordinates: "we ... view mass as on the same footing as time and space ... " [20] and "This means that the role of the 4D uncharged mass is played in 5D geometry by the extra coordinate." We find this approach to be irrational. In this context, it poses a difficulty to generalize the energy-momentum-mass tensor in four-dimensional space to the tensor in five dimensions. The fifth coordinate, mass, can be utilized, but not in the coordinate space. The mass of the particle should be viewed as an additional value to the energy and other three momentum components. The fifth coordinate in coordinate space should not have a value related to the mass. It was hard to draw connections with actual experiments as a result of the assignment of mass as a fifth coordinate in addition to time and space.

The physical meaning of the fifth coordinate is action. This value is constant under the usual Lorentz transformations in M, but it changes when the transformations in the extended space G(T, X, Y, Z, S) are used. From a physical point of view, our expansion means that processes in which the rest mass of the particles changes are acceptable now. Lorentz transformations in the 4D Minkowski space M(T; X, Y, Z) in the planes (T, X), (T, Y), (T, Z) allow changing the energy and momentum of a particle in the conjugate space of the expanded 4D energy-momentum space M∗(E; Px, Py; Pz). In the ESM, gravity and electromagnetism are combined in one field, and it is possible to construct a 5×5 energy-momentum-mass tensor. Recently, Overduin and Henry [32] proposed the same idea of considering the fifth coordinate.

In ESM, the motion of a particle in a 5-dimensional cone

$$(ct)^2 - x^2 - y^2 - z^2 - s^2 = 0 \tag{6}$$

is considered. The parameters t, x, y, z, s are coordinates of a point in extended space G(1,4). The Minkowski space M(1,3) enters it as a subspace. In the extended space G(1,4) the particle 4-vector of energy and momentum is padded to a five-dimensional vector to a 5-vector

$$\overline{p} = \left(\frac{E}{c}, p_x, p_y, p_z, mc\right) \tag{7}$$

where E is energy, and p_x, p_y, p_z are momenta.

For a free particle, the components of this vector satisfy

$$\frac{E^2}{c^2} - p_x^2 - p_y^2 - p_z^2 - m^2c^2 = 0 \tag{8}$$

The interval s in Minkowski space serves as the fifth coordinate in the extended space G(1,4). The variations of mass m correspond to variations of the interval s. Let us now explain the fifth coordinate physical meaning in the ESM. For this purpose, we use the expression for the action S of a free particle [20, 21].

$$S = \int_a^b L ds \tag{9}$$

Here L is the Lagrangian of a particle. The integral is taken along the world line between two given events—the position of the particle at the beginning and end points of the Minkowski space.

Since the mass of the photon in STR is $m = 0$, massless fields in the extended space G(1,4) are mapped to a 5-vector

$$\overline{p}_f = \left(\frac{\omega}{c}, \frac{\omega}{c}\vec{k}, 0\right) \tag{10}$$

We assume that a photon corresponds to a plane wave, which moves in M(1,3) with a speed c in the direction given by the vector $\vec{k}$. We want to consider a broader class of processes that can change mass. The energy-momentum-mass 5-vector (11) characterizes a particle for which all the parameters, energy, momentum, and mass are variables. The corresponding changes of these values can be described using transformations of the extended space G(1,4), given by the hyperbolic rotations on an angle ϕ_{TS} in the plane (TS)

$$\frac{E'}{c} = \frac{E}{c}\cosh\phi_{TS} + p_s \sinh\phi_{TS}, \quad P' = P, p'_s = p_s \cosh\phi_{TS} + \frac{E}{c}\sinh\phi_{TS} \tag{11}$$

and in the plane (XS)

$$\frac{E'}{c} = \frac{E}{c}, \quad P' = P\cosh\phi_{XS} + p_s \sinh\phi_{XS}\, p'_s = p_s \cosh\phi_{XS} + P\sinh\phi_{XS} \tag{12}$$

The extended space G(1,4) can be viewed as a set of Minkowski spaces with the parameter n, which we conditionally refer to as the refractive index. We made this parameterization option because the physical meaning of our model depends on the photon's movement and its velocity variations. We believe that the refractive index n of any subspace M(1,3) of space G(1,4) defines these subspaces. From the point of view of ESM, the transition from a medium with one refractive index n_1 to a medium with another refractive index n_2 can be interpreted as the movement along the fifth coordinate of the expanded space. In contrary to the usual relativistic mechanics, we now suppose that the mass m is also a variable, and it can vary at motion of a particle on the cone (6). In the ESM model when a particle enters an area of space with a non-zero density of matter or field, its mass changes. In such areas, the speed of light is reduced, and these areas can be characterized by the refractive index of a medium n. This parameter relates the speeds of light in the vacuum and in the medium, which is $v = c/n$. For example, the refractive index of a gravitational field that is described by the Schwarzchild solution [33] reads

$$n(r) = \left(1 - \frac{\alpha}{r_g}\right)^{-1} \tag{13}$$

where r is a distance from gravity center. The values of the components of photon 5-vector (7) corresponding to this field are found using (TS) rotation (11) and will be transformed as follows:

$$\left(\frac{\hbar\omega}{c}, \frac{\hbar\omega}{c}, 0\right) \to \left(\frac{\hbar\omega}{c}\cosh\phi, \frac{\hbar\omega}{c}, \frac{\hbar\omega}{c}\sinh\phi\right) \\ = \left(\frac{\hbar\omega}{c}n, \frac{\hbar\omega}{c}, \frac{\hbar\omega}{c}\sqrt{n^2-1}\right) \tag{14}$$

Electromagnetic and gravitational fields combined in a single gravitational-electromagnetic field. This is because in the ESM, electromagnetic interaction causes the formation and changing of fields and particle mass. Since mass is assumed to be variable in the ESM from the beginning, it is possible to describe new processes that the STR framework is unable to explain properly. This arises from the fact that in ESM, gravitational interaction of elementary particles occurs naturally.

3. Plane wave localization and mass appearance in the ESM

In the extended space G(1,4), the potentials of the field combining electromagnetism and gravity are determined by the equation [15]:

$$\Pi_{(5)}A_0 = -4\pi e, \quad \Pi_{(5)}\vec{A} = -\frac{4\pi}{c}\vec{j}, \quad \Pi_{(5)}A_s = -\frac{4\pi}{c}j_s$$

with

$$\Pi_{(5)} = \frac{\partial^2}{\partial s^2} + \frac{\partial^2}{\partial x^2} + \frac{\partial^2}{\partial y^2} + \frac{\partial^2}{\partial z^2} - \frac{1}{c^2}\frac{\partial^2}{\partial t^2} \tag{15}$$

and the charge of a particle e.

On the right-hand side of the equations in system (15) are the components of the five-dimensional current vector $\overline{\rho}$. This vector is a generalization of the four-dimensional current vector $\tilde{\rho}$, which in traditional four-dimensional electrodynamics is

$$\tilde{\rho} = \left(e, \vec{j}\right) = \left(\frac{e_0 c}{\sqrt{1-\beta^2}}, \frac{e_0\vec{v}}{\sqrt{1-\beta^2}}\right); \; \tilde{\rho}^2 = e_0^2 c^2 \tag{16}$$

where e is the charge of a particle.

In order to get $\overline{\rho}$, we assign an additional component to the vector (16):

$$\overline{\rho} = \left(\rho, \vec{j}, j_s\right) \tag{17}$$

This is an isotropic vector: $\overline{\rho}^2 = 0$. We will consider an additional component as an analog of the momentum of a charged sphere moving at low speed [34] in an extra dimension

$$j_s = \frac{4}{3}\kappa\frac{e_0^2}{8\pi c\varepsilon_0 a}\frac{ds}{d\mu} \tag{18}$$

where ε_0 is the vacuum permittivity, a is the sphere radius, μ is an arbitrary affine parameter along the path, and κ is a constant having the dimension of the ratio of charge to mass. Assuming that e_0 and $a = r_e$ are the charge and radius of the electron and the constant κ is equal to the ratio of its charge to mass: $\kappa = e_0/m_e$, we get

$$j_s = \frac{4}{3}\frac{e_0^3}{8\pi c\varepsilon_0 r_e m_e}\frac{ds}{d\mu} \tag{19}$$

The electromagnetic field is defined by the system of the first four Eq. (15), while the fifth equation describes the scalar gravitational field. A single electrogravity field results from the combination of these two fields if their values depend on the variable s. For such fields, Lienard-Wichert potentials were found in [35]. When the variables included in potentials have no dependence on the variable s, it splits into two independent subsystems.

From electrostatics theory [36], the potential energy of a uniformly charged sphere is given by

$$E_{pot} = \frac{e_0^2}{8\pi c\varepsilon_0 a} \tag{20}$$

When the fifth equation describes the gravitational field created by the electrons charge, their gravity density j_s, Eq. (19), will not be equivalent to, their potential energy and will depend on the speed of movement in the extra dimension $ds/d\mu$. Accordingly, the radius of an electron may differ from its classical radius $r_{e\ cl} = 2.8 \bullet 10^{-15}\ m$, determined from Eq. (20). This is confirmed by observation of a single electron in a Penning trap [37], which suggests the upper limit of the particle's radius to be $10^{-22}\ m$.

Now let us look how a charged particle and a plane electromagnetic wave interact. We take it as given that a plane wave is an object in empty space that fills this infinite space. The following equations are used to find the field potentials without sources in the extended space G(1,4):

$$\Pi_{(5)}A_0 = 0, \quad \Pi_{(5)}\vec{A} = 0, \quad \Pi_{(5)}A_s = 0 \tag{21}$$

Let us consider the equation

$$\frac{\partial^2}{\partial s^2}u + \frac{\partial^2}{\partial x^2}u + \frac{\partial^2}{\partial y^2}u + \frac{\partial^2}{\partial z^2}u - \frac{1}{c^2}\frac{\partial^2}{\partial t^2}u = 0 \tag{22}$$

We are looking for its solutions in the form

$$U(s,x,y,z,t) = u(s,x,y,z)\mathrm{e}^{-iks}\mathrm{e}^{i\omega t}, \quad k = \frac{2\pi}{\lambda} \tag{23}$$

We assume that the function $u(s,x,y,z)$ varies slowly over the variable s, compared with the variables x,y,z, so that the second derivative $(\partial^2/\partial s^2)u$ can be neglected. Now we get the equation

$$\frac{\partial^2}{\partial x^2}u + \frac{\partial^2}{\partial y^2}u + \frac{\partial^2}{\partial z^2}u - 2ik\frac{\partial}{\partial s}u = 0 \tag{24}$$

The neglect of the second derivative and presentation of Eq. (22) in the form (24) are similar to the searching a solution for the optical wave propagating shape in a laser along the z-axis [38]. Eq. (24) solution has the form of a three-dimensional Gaussian beam

$$u = u_0\left(\frac{w_0}{w}\right)^{3/2} exp\left[-i(ks+\varphi) - \left(x^2+y^2+z^2\right)\left(\frac{1}{w^2}+\frac{ik}{2R}\right)\right] \quad (25)$$

Here w_0 is the radius of the "neck" of the beam, that is, its minimum width at the point $s = 0$. The value of $w^2 = w_0^2\left[1+\left(2s/kw_0^2\right)^2\right.$ is the diameter of the beam at the point z. The radius of curvature of the beam wavefront is $R(z) = z\left[1+\left(kw_0^2/2s\right)^2\right]$.

For $s \to \infty$, the radius and the beam width also tend to infinity. Solution (25) corresponds to a plane wave. If $s \to 0$, the plane wave is localized in a volume that looks like a ball with radius $r = w_0$. Process of localization takes place without changing the energy. It is described by orthogonal rotations in the planes (SX), (SY), (SZ). The square of the wave modulus (25) reads as

$$|u|^2 = |u_0|^2\left(\frac{w_0}{w}\right)^3 \exp\left[-\left(x^2+y^3+z^2\right)\left(\frac{2}{w^2}\right)\right] \quad (26)$$

As we can see, as the wave's localization (25) diminishes, it grows and achieves its maximum value at $s = 0$.

This degree of localization is not, however, really reached. The presence of a charged particle in space causes the localization of a plane wave. The charged particle effects at the wave field, but the field also has an impact on the particle. Since a charged particle's mass distribution is defined by δ function, we suppose that it is concentrated at a single point in empty space (refractive index $n = 1$), and that this delta function is the free particle wave function. The-function starts to change into a Gaussian function as soon as such a charged particle reaches the plane-wave field.

We see that as s decreases, the localization of the wave (25) increases and reaches its maximum value at $s = 0$. However, this degree of localization is not really achieved. The fact is that the process of localization of a plane wave is generated by the presence of a charged particle in space. The wave field is affected by the charged particle, but the particle itself is affected by the field. We assume that in empty space (refractive index $n = 1$), a charged particle is concentrated at a point, that is, its mass distribution described by δ is a function, and we consider this delta function as the free particle wave function. When such a charged particle enters the plane-wave field, the δ-function begins to transform into a Gaussian function

$$|v_s(x,y,z)|^2 = K|v_0|^2\left(\frac{1}{s^2\pi}\right)^3 \exp\left[-\left(x^2+y^3+z^2\right)\left(\frac{2}{s^2}\right)\right] \quad (27)$$

This expression is structurally similar to the solution of the differential heat equation describing the temperature distribution. For an infinite body with an instantaneous point source at the origin, the temperature distribution has the following form:

$$T(x,y,z,t) = \frac{Q}{\varsigma\rho(4\pi at)^{3/2}} \exp\left(-\frac{x^2+y^2+z^2}{4at}\right) \quad (28)$$

Here T is the temperature at time t in coordinates x, y, z; Q is the heat emitted at the time $t = 0$ at the origin; t is the time elapsed since the introduction of heat; a is the thermal diffusivity, ρ is the density of the body, and ς is its specific heat. Eq. (28) is the fundamental solution of the heat equation under the action of an instantaneous point source in an infinite body.

The localization of a plane wave and the "swelling" of a massive particle are both consistent processes. Our assumption is that their localizations must match in order for the field and particle to interact. We get an expression for the value s_0 that defines the lowest value of the plane-wave localization and the maximum value of the point particle swelling by comparing Eqs. (26) and (27):

$$s_0^2 = \frac{k^2 w_0^4}{k^2 w_0^2 - 4} = \frac{\pi^2 w_0^4}{\pi^2 w_0^2 - \lambda^2} \tag{29}$$

This imposes restrictions on the dynamics of interaction of the field with the particle.

4. Equations of Lagrange mechanics

In GRT, a four-dimensional pseudo-Riemannian space-time with coordinates x^i and metric coefficients g_{ij} is considered, the interval in which is written in the form

$$ds^2 = g_{ij} dx^i dx^j \tag{30}$$

The 4-velocity vector of the particle is denoted as $u^i = dx^i / d\mu$, where μ is the variable parameter. We obtain the equations of its dynamics in general form.

The particle Lagrangian corresponds to the covariant generalized momenta

$$p_i = \frac{\partial L}{\partial u^i} \tag{31}$$

and generalized forces

$$F_i = \frac{\partial L}{\partial x^i}. \tag{32}$$

The particle motion is determined by Hamilton's principle of stationary action $\delta S = 0$ at

$$S = \int_{\mu_0}^{\mu_1} L d\mu \tag{33}$$

where μ_0, μ_1 are the values of the parameter at the points that are connected by the desired trajectory of motion. The extremum condition leads to the Euler-Lagrange equations

$$\frac{d}{d\mu} \frac{\partial L}{\partial u^\lambda} - \frac{\partial L}{\partial x^\lambda} = 0 \tag{34}$$

With generalized momenta (2.2) and forces (2.3), these equations are rewritten in the form

$$\frac{dp_\lambda}{d\mu} - F_\lambda = 0 \tag{35}$$

The Lagrangian is chosen [21] so that contravariant momenta bind to the physical energy and momentum of the particle

$$p^j = g^{j\lambda} p_\lambda \tag{36}$$

and the gravitational force acting on it is mapped to associated with (32) vector

$$F^l = g^{l\lambda} F_\lambda \tag{37}$$

Passing to them in Eq. (35), we find

$$g_{\lambda i} F^i = g_{\lambda i} \frac{dp^i}{d\mu} + \frac{\partial g_{\lambda i}}{\partial x^l} u^l p^i \tag{38}$$

Multiplying these equations by $g^{k\lambda}$ and summing over the twice occurring index λ, we obtain.

$$F^k = \frac{dp^k}{d\mu} + g^{k\lambda} \frac{\partial g_{\lambda i}}{\partial x^l} u^l p^i. \tag{39}$$

The presence of the second term on the right side reflects that in the gravitational field not only the 4-momentum of matter, but the 4-momentum of matter together with the gravitational field is stored [18]. Its components express the rate of change of the energy and momentum acquired by the gravitational field when a particle moves in it

$$\frac{d\overset{\leftrightarrow}{p}^k}{d\mu} = g^{k\lambda} \frac{\partial g_{\lambda i}}{\partial x^l} u^l p^i. \tag{40}$$

Integration of this quantity over gives the energy and momentum received by the gravitational field at a certain interval of its trajectory. As a result, Eq. (39) can be written in the form

$$F^k = \frac{dp^k}{d\mu} + \frac{d\overset{\leftrightarrow}{p}^k}{d\mu}. \tag{41}$$

It follows from the laws of conservation of energy and momentum that the force acting on a particle is equal in magnitude and opposite in sign to the force acting by the source of gravity from the side of the particle. This is equivalent to fulfilling Newton's third law.

5. Variational principle of the energy stationary integral for the photon motion

To determine the dynamics of a photon in a gravitational field, we will use principle of the energy stationary integral [21–24]. Interval in pseudo-Riemannian space-time with metric coefficients $\tilde{g}_{ij}$:

$$ds^2 = \tilde{g}_{ij} dx^i dx^j \tag{42}$$

after substitutions $\tilde{g}_{11} = \rho^2 g_{11}, \tilde{g}_{1p} = \rho g_{1p}, \tilde{g}_{pq} = g_{pq}$ at $p, q = 2, 3, 4$ is rewritten as

$$ds^2 = \rho^2 g_{11} dx^{12} + 2\rho g_{1p} dx^1 dx^p + g_{pq} dx^p dx^q. \tag{43}$$

The condition $ds = 0$ corresponds to the motion of light. With $g_{11} \neq 0$, the variable ρ is given by the expression

$$\rho = \frac{-g_{1p} u^p + \sigma \sqrt{\left(g_{1p} g_{1q} - g_{11} g_{pq}\right) u^p u^q}}{g_{11} u^1}, \tag{44}$$

where σ takes the values ± 1 and 4-velocities u^i are determined provided that μ is an affine parameter. Further, we will consider variations near $\rho = 1$, to which the equality $\tilde{g}_{ij} = g_{ij}$ corresponds. If $g_{11} = 0$ and condition $g_{1p} \neq 0$ is satisfied for at least one ρ, then it turns out

$$\rho = -\frac{g_{pq} u^p u^q}{2 g_{1k} u^1 u^k}, \tag{45}$$

where k takes on the values (2)–(4).

The Lagrangian of a freely moving particle is chosen as

$$L = -\rho. \tag{46}$$

For both values (44), (45), the covariant generalized momenta (31) and forces (32) take the form

$$p_\lambda = \frac{u_\lambda}{u^1 u_1}, \tag{47}$$

$$F_\lambda = \frac{1}{2 u_1 u^1} \frac{\partial g_{ij}}{\partial x^\lambda} u^i u^j. \tag{48}$$

The chosen Lagrangian corresponds to the ratio

$$\rho = u^\lambda \frac{\partial L}{\partial u^\lambda} - L \tag{49}$$

being the integral of motion [39] and, accordingly, ρ will be the energy of the system combining the light-like particle and the gravitational field given by the metric (30).

The equations of motion are found from Hamilton's principle of stationary action (33), which, in view of (46), can be written in the form

$$S = -\int_{\mu_0}^{\mu_1} \rho \, d\mu. \tag{50}$$

The energy ρ is non-zero, its variations leave the interval light-like. The equations of motion will be Euler-Lagrange Eq. (35). The principle of the energy stationary integral for the photon motion is consistent [23, 24] with the generalized Fermat's principle [40], and the resulting curves are null geodesics.

The contravariant vector of generalized momenta is written as

$$p^{\lambda} = \frac{1}{u^1 u_1} u^{\lambda}. \tag{51}$$

Physical energy and momenta of photon with frequency ν in Minkowski space-time with affine parameter $\mu = ct$ form contravariant 4-vector of momenta $\pi^i = (\hbar\nu/c)u^i$. For arbitrary affine parameter, it is rewritten as

$$\pi^i = \frac{\hbar\nu}{c}\frac{u^i}{u^1}. \tag{52}$$

And in pseudo-Riemannian space-time, similar energy and momenta of the photon will be put in line with the components of the contravariant vector of momenta. The photon frequency in coordinate frame is given by

$$\nu = \frac{\nu_0}{u_1}, \tag{53}$$

where ν_0 is a certain fixed value of its frequency. Comparing Eqs. (51) and (52) we obtain

$$\pi^i = \frac{h\nu_0}{c} p^i. \tag{54}$$

The Lagrangian (46) corresponds to a particle with unit energy. For a photon, it is

$$L_{ph} = \frac{h\nu_0}{c} L \tag{55}$$

and the gravitational forces acting on photon

$$Q^l = h\nu_0 F^l \tag{56}$$

are assigned to the components of the associated vector of generalized forces

$$F^l = g^{l\lambda} \frac{1}{2u_1 u^1} \frac{\partial g_{ij}}{\partial x^{\lambda}} u^i u^j. \tag{57}$$

6. Photon dynamics in the Schwarzschild field

Let us consider the dynamics of a light-like particle in a static centrally symmetric gravitational field described in spherical coordinates $(t, r, \theta, \phi)$by the Schwarzschild metric

$$ds^2 = c^2\left(1 - \frac{r_g}{r}\right)dt^2 - \left(1 - \frac{r_g}{r}\right)^{-1} dr^2 - r^2(d\theta^2 + sin^2\theta d\varphi^2), \tag{58}$$

In plane $\theta = \pi/2$, the energy-momentum vector of a photon (54) moving along an open trajectory [21–24] is as follows:

$$\pi^i = \left(\frac{h\nu_0}{1 - \frac{r_g}{r}}, \quad \pm h\nu_0 \sqrt{1 - \frac{C^2}{r^2}\left(1 - \frac{r_g}{r}\right)}, \quad 0, \quad \frac{h\nu_0 C}{r^2} \right), \tag{59}$$

where C is constant, and ν_0 is a photon frequency away from the center of gravity. In radial motion, the components of this vector coincide with the first components of the vector (14) obtained by (TS)-rotation according to the ESM.

The only non-zero component of the associated vector of generalized forces (57) is

$$F^2 = -\frac{r_g}{r^2} + \frac{C^2}{r^3}\left(1 - \frac{r_g}{r}\right)\left(1 - \frac{r_g}{2r}\right). \tag{60}$$

With radial motion ($C = 0$), it is equal to

$$F^2 = -\frac{r_g}{r^2}, \tag{61}$$

coinciding with the doubled force acting on the particle in Newtonian gravity. In view of (56), it corresponds to the passive gravitational mass of the photon

$$m_p^{ph} = \frac{2h\nu_0}{c^2} \tag{62}$$

This result is consistent with a thought experiment on "weighing" a photon [41], in which it performs periodic motion in the vertical direction between two horizontal reflecting surfaces.

Considering the non-radial motion, in order to avoid the appearance of a fictitious component of momenta and force due to the sphericity of the coordinate system, we use the Schwarzschild metric in rectangular coordinates. It can be accessed using the transformation

$$r = \left(1 + \frac{r_g}{4\bar{r}}\right)^2 \bar{r}, \tag{63}$$

$$x = \bar{r}\cos\theta\cos\varphi, \quad y = \bar{r}\cos\theta\sin\varphi, \quad z = \bar{r}\sin\theta \tag{64}$$

of metric (58), which yields

$$ds^2 = c^2\left(\frac{1 - \frac{r_g}{4\bar{r}}}{1 + \frac{r_g}{4\bar{r}}}\right)^2 dt^2 - \left(1 + \frac{r_g}{4\bar{r}}\right)^4 \left(dx^2 + dy^2 + dz^2\right). \tag{65}$$

The motion in the plane $z = 0$ is studied and the force acting on a light-like particle at a point $(ct, x, 0, 0)$.

The single non-zero component of the force vector [21–24] is

$$F^2_{rect} = -\frac{r_g\left(1 - \frac{r_g}{8\bar{r}}\right)}{\bar{r}^2\left(1 + \frac{r_g}{4\bar{r}}\right)^5\left(1 - \frac{r_g}{4\bar{r}}\right)}, \tag{66}$$

which taking into account transformation (63) can be rewritten as

$$F^2_{rect} = -\frac{r_g\left(1 - \frac{r_g}{8r}\right)}{r^2\left(1 - \frac{r_g{}^2}{16r^2}\right)}. \tag{67}$$

The generalized force acting on a photon does not depend on the direction of its motion. This expression differs from the formula (61) corresponding to radial motion in spherical coordinates, which is a consequence of the non-covariance of the vector F^l. However, in the limit of weak gravity ($r \gg r_g$), these expressions converge asymptotically and give Newton's law of gravitation with a passive gravitational mass of a photon (62) equal to twice the mass of a material particle of equivalent energy.

The gravitational field of the electromagnetic radiation flux is determined from the solution of the Einstein equations

$$R_{ij} - \frac{1}{2}g_{ij}R = \chi T_{ij} \tag{68}$$

with Ricci tensor R_{ij} and $\chi = \frac{8\pi G}{c^4}$ for the electromagnetic field energy-momentum tensor

$$T^{EM}_{ij} = \frac{1}{4}g_{ij}F_{kl}F^{kl} - F^k_i F_{jk}, \tag{69}$$

where F_{ij} is the electromagnetic field tensor. In case of weak gravity, it follows from analysis of acceleration of material particle that active gravitational mass of light beam or light packet is twice as much as similar mass of a rod of equivalent energy [26, 27, 42]. The equality of the active and passive gravitational masses of a photon means the fulfillment of Newton's third law in the gravitational interaction of light and material particles and the laws of conservation of energy and momentum.

7. Electron positron pair production, gravitons

The expansion of the Birkhoff theorem to a sphere with equally distributed electrons and positrons provided the justification for applying the energy conservation law to a gravitational field source at the annihilation reaction [26]. This was done under the assumption that the gravitational mass of the sphere would not change because of electron-positron annihilation before the particles left it. Energy conservation law for the source of the gravitational field applies to any pair of annihilating particles if this condition is fulfilled for the entire sphere. Furthermore, due to the double gravitational mass of the photon compared with the total mass of the electron and positron $2m_{e^-}$, the annihilation reaction results in the appearance of particles g^- with a negative gravitational mass

$$m^{gr}_{g^-} = -m_{e^-}, \tag{70}$$

dissipating the negative energy as a gravitational source.

Annihilation process in this case looks as follows.

$$e^- + e^+ \to 2\gamma + 2g^- \tag{71}$$

There is no energy for particles g^- specified by non-gravitational interactions. This is because the total electron and positron energy is equal to the gamma quantum energy produced. These particles have no kinetic momentum. As a result, it is impossible to detect them using conventional particle registration techniques (such as a bubble chamber). However, when a ray of light passes through the negative energy region, a focusing effect can occur, as opposed to focusing by gravitational lensing [43].

Conditions for the formation of electron-positron pairs are produced by high-energy gamma radiation interactions (>1022 MeV) with matter. A "gravitational charge" particle, designated g^+, appears as a result of the inverse annihilation reaction

$$2\gamma \to e^- + e^+ + 2g^+. \tag{72}$$

Appearing in addition to the electron and positron particles are opposite to g^- in gravitational mass. It occurs with extracting pairs g^+, g^- from a vacuum. Particle g^- is immediately absorbed, producing an electron and a positron, leaving g^+ with positive gravitational mass. Particles g^- are immediately absorbed, producing an electron and a positron and leaving g^+ with a positive gravitational mass.

We consider a model in which the boson-like particles g^- and g^+ have a rest mass of 0. These particles are assumed to have a spin of 2 and an electric charge of 0. The graviton, a hypothetical quantum of gravitational radiation, holds the property of particle g^+ [28]. Fermi Gamma-ray Space Telescope [44] detects photons with an energy, sufficient for reaction (72), in pulsar jets, such as Cygnus X-3 [45], gamma-ray bursts from blazars [46], and cosmic ray generation in supernova remnants [47].

Gravitational mass of body is less than the sum of individual gravitational masses its constituent elements [48]. The gravastar [49] with negative mass component with half the mass two times less than the mass obtained by integrating the spatial volume is considered in [17]. In this paper, the gravitational mass defect is inspected as a result of the negative binding energy presence. This case reflects to the ratio between the electron mass and gravitational mass of the particle g^- released during annihilation. Such condition corresponds to the negative internal pressure and positive pressure on the shell. This model is a gravastar with inside approaching a de Sitter interior vacuum with constant density and pressure, having a singularity near the shell [49].

8. Gravastar with constant pressure

The general static, spherically symmetric line element in Schwarzschild coordinates is

$$ds^2 = f(r)\,dt^2 - \frac{dr^2}{h(r)} - r^2\left(d\theta^2 + \sin^2\theta\, d\phi^2\right) \tag{73}$$

with metric functions $f(r)$ and $h(r)$. The stress-energy tensor of a static, spherically symmetric distribution of matter with density q and isotropic pressure p is described (in units in which $c = 1$) by the diagonal matrix

$$T_i^i = diag(q, -p, -p, -p). \tag{74}$$

Metric functions are sought from the Einstein Eq. (68), which reduced [49] to

$$G^t{}_t = \frac{1}{r^2}\frac{d}{dr}[r(1-h)] = 8\pi G\rho, \tag{75}$$

$$G^r{}_r = -\frac{h}{rf}\frac{df}{dr} + \frac{1}{r^2}(1-h) = -8\pi Gp. \tag{76}$$

These equations have a solution [50]:

$$h(r) = 1 - \frac{8\pi G}{3}\rho r^2, \tag{77}$$

$$f(r) = const \tag{78}$$

with constant density and pressure obeying the relation

$$p = -\frac{1}{3}\rho. \tag{79}$$

Metric functions f and h must match the exterior Schwarzschild solution (58) in vacuum

$$f_{ext}(r) = h_{ext}(r) = 1 - \frac{r_g}{r}, \quad r_g = \frac{8\pi G}{3}\rho R^3, \quad r \geq R, \tag{80}$$

where R is radius of matter distribution. Thus, we have the boundary conditions

$$f(R) = h(R) = 1 - \frac{r_g}{R} = 1 - \frac{8\pi G}{3}\rho R^2, \tag{81}$$

and since function f is constant inside the sphere (78), it will have the value

$$f(r) = 1 - \frac{8\pi G}{3}\rho R^2 \tag{82}$$

The space-time inside the sphere is the following:

$$ds^2 = \left(1 - \frac{8\pi G}{3}\rho R^2\right)dt^2 - \frac{dr^2}{1 - \frac{8\pi G}{3}\rho r^2} - r^2\left(d\theta^2 + \sin^2\theta\, d\phi^2\right). \tag{83}$$

Equation of state (79) characterizes the vacuum pressure that balances the pressure of the gravitational field in a uniform sphere [21, 50, 51] or, in case of electromagnetic field, pressure of it [27]. This gravastar model will have the same dependence between its proper and negative mass components on radius and density as the gravastar with a de Sitter interior vacuum [28, 49].

9. Conclusions

The generalization of Einstein's special theory of relativity on 5-dimentional space is considered, in which as fifth coordinate additional coordinate is identified with the interval of a particle. We obtained a 5D generalization of the 4D current vector under the assumption that the mass of an electron is equivalent to a component of its momentum dependent on its velocity in an additional dimension. As follows from this approach, the radius of an electron can be substantially less than its classical radius.

A point charged particle's field is entered by a plane electromagnetic wave. It was found that such a wave can be described by potentials in a five-dimensional extended space. It is considered on how such a wave could interact with a charged particle. We can calculate the field strengths and determine their energy-momentum-mass tensor using the explicit form of these potentials.

The dynamics of particles in curvilinear space-time is considered using Lagrange mechanics. A correspondence is established between the physical energy and momentum of a particle, determined from non-gravitational interactions, and the contravariant vector of generalized momenta. The obtained dynamic equations include the rate of change of the energy-momentum vector, the components of which express the energy and momentum acquired by the gravitational field when a particle moves in it. This vector is an analog of the pseudotensor used in conservation laws in tensor form when considering the dynamics of an individual particle.

By choosing the Lagrangian of a photon corresponding to the principle of the energy stationary integral, a vector of forces acting on it in the Schwarzschild field is obtained. Although these generalized forces are not covariant quantities, in the limit of weak gravity, they express the Newtonian law of gravity with a passive mass of particles corresponding to the active gravitational mass of moving point bodies and a light beam. The passive gravitational mass of a photon does not depend on the direction of its motion. Coinciding with its active gravitational mass when interacting with a material particle, it is equal to twice the mass of a material particle having an energy equivalent to a photon. When a particle moves in a gravitational field, the non-covariance of the generalized forces vector and the vector composed of the rates of energy and momentum transfer to it has the same nature as the non-covariance of the gravitational field energy-momentum pseudo-tensor, with the help of which experimentally confirmed changes in the circular orbits of two bodies moving around a common center were calculated as a result of energy loss caused by the radiation of gravitational waves.

With twice the gravitational photon mass compared with a material particle of equivalent energy, the application of Birkhoff's theorem to a sphere full of annihilating electrons and positrons leads to the following conjecture: during electron-positron annihilation, particles g^- with a negative gravitational mass are released in addition to gamma quanta. The opposite "gravitational charge" particle g^+ will appear as a result of the photon conversion event into an electron-positron pair. With a spin of 2 and no electric charge, these hypothetical particles are categorized as bosons. They are identified as gravitons, being the localization of elementary gravitational waves. Gamma-ray blazar outbursts, the jets of pulsars like Cygnus X-3, and cosmic ray emissions in supernova remnants contain photons with enough energy for their appearance.

We proposed a gravastar model with constant pressure and no singularity. The gravitational mass defect is explained by the presence of negative binding energy. This model allows for a relation between the gravitational mass and the negative component of mass corresponding to the relation between the mass of electron and particle g^-. The equation of state in the inner region is such that the pressure of the vacuum balances the pressure of the gravitational or electromagnetic fields.

Author details

Dmitry Yu Tsipenyuk[1,2,3*] and Wladimir B. Belayev[4]

1 Moscow Polytechnic University, Moscow, Russia

2 Moscow Institute of Physics and Technology, Moscow Region, Russia

3 Russian Institute for Scientific and Technical Information of the Russian Academy of Sciences, Moscow, Russia

4 Khlopin Radium Institute, Saint-Petersburg, Russia

*Address all correspondence to: dimat777@list.ru

References

[1] Klein F. Über neuere englische arbeiten zur gesammelte matematishe abhandlungen. Zeitschrift für Angewandte Mathematik und Physik. 1901;**46**:376

[2] Einstein A. Die Grundlage der allgemeinen relativitätstheorie. Annals of Physics. 1916;**49**:769-822. DOI: 10.1002/andp.19163540702

[3] Kaluza T. Zum unitätsproblem der physik. Sitzungsberichte der Preussischen Akademie der Wissenschaften. 1921;7:966-972. DOI: 10.1142/S0218271818700017

[4] Klein O. Quantentheorie und fünfdimensionale relativitätstheorie. Zeitschrift für Physik. 1926;**37**(12):895-906. DOI: 10.1007/BF01397481

[5] Mandel H. Über den zusammenbang zwischen der Einsteinschen theorie des fern. Parallelismus und der fünfdimensionalen fieldtheorie. Zeitschrift für Physik. 1926;**39**:136-145

[6] Fock VA. Über die invariante form der wellen und der bewegungsgleichungen für einen geladenen massenpunkt. Zeitschrift für Physik. 1926;**39**:226-232. DOI: 10.1007/BF01321989

[7] Einstein A, Bergmann P. Generalization of Kaluza's theory of electricity. Annals of Mathematics. 1938; **39**(3):683-701. DOI: 10.2307/1968642

[8] Rumer YB. Issledovaniya po Pyatimernoy Optike [Studies on Five-Dimensional Optics]. Moscow: URSS; 2010. p. 152

[9] Konopleva NP, Popov VN. Kalibrovochnye Polya [Gauge Fields]. Moscow: Atomizdat; 1972. p. 240

[10] Rubakov VA. Klassicheskie Kalibrovochnye Polya [Classical Gauge Felds]. Moscow: URSS; 1999. p. 336

[11] Stanyukovich KP. Gravitatsionnoe Pole i Elementarnye Chastitsy [Gravitational Field and Elementary Particles]. Moscow: Nauka; 1965. p. 312

[12] Markov MA. Elementary particles of maximally large masses (quarks and maximons). Journal of Experimental and Theoretical Physics. 1967;**24**:584-592

[13] Landau LD. On the radius of elementary particles. Journal of Experimental and Theoretical Physics. 1940;**10**:718-722

[14] Tsipenyuk DY, Andreev VA. Structure of extended space. Bulletin of the Lebedev Physics Institute. 2000;**6**: 23-34. DOI: 10.48550/arXiv.gr-qc/0106093

[15] Andreev VA, Tsipenyuk DY. The 5-dimensional model for electromagnetism and gravity. Natural Science. 2014;**6**: 248-253. DOI: 10.4236/ns.2014.64028

[16] Andreev VA, Tsipenyuk DY. The mass and size of photons in the 5-dimensional extended space model. Journal of Modern Physics. 2016;**7**: 1308-1315. DOI: 10.4236/jmp.2016.711116

[17] Andreev VA, Tsipenyuk DY. The problem of introducing finite size and variable mass of the photon. Inzhenernaya Fisika. 2017;**5**:17-28. Available from: http://infiz.tgizd.ru/en/arhiv/16341

[18] Landau LD, Lifshitz EM. The Classical Theory of Fields. 3rd ed. Singapore: World Scientific; 1994

[19] Weinberg S. Gravitation and Cosmology: Principles and Applications of the General Theory of Relativity. New York: Wiley; 2013. p. 688

[20] Ritus VI. Lagrange equations of motion of particles and photons in the Schwarzschild field. Physics-Uspekhi. 2015;**58**:1118-1123. DOI: 10.3367/UFNe.0185.201511h.1229

[21] Belyaev VB. Dinamika v Obschei Teorii Otnositel'nosti: Variacionnye Metody [the Dynamics in General Relativity Theory: Variational Methods]. Moscow: URSS; 2017. p. 216

[22] Tsipenyuk DY, Belyaev VB. Extended space model is consistent with the photon dynamics in the gravitational field. Journal of Physics: Conference Series. 2019;**1251**:012048. DOI: 10.1088/1742-6596/1251/1/012048

[23] Tsipenyuk DY, Belyaev VB. Photon dynamics in the gravitational field in 4D and its 5D extension. Romanian Reports in Physics. 2019;**71**:109

[24] Tsipenyuk DY, Belyaev VB. Bubble structures in microphysical objects in 5-D extended space model. Radioelectronics Nanosystems Information Techologies. 2019;**11**:249. DOI: 10.17725/rensit.2019.11.249

[25] Fock V. The Theory of Space Time and Gravitation. 2nd ed. Amsterdam: Pergamon; 1964. p. 460

[26] Tolman RC, Ehrenfest P, Podolsky B. On the gravitational field produced by light. Physics Review. 1931; **37**:602-615. DOI: 10.1103/PhysRev.37.602

[27] Tolman RC. Relativity, Thermodynamics and Cosmology. Oxford: Clarendon Press; 1969. p. 502

[28] Tsipenyuk DY, Belayev WB. Astrophysical and microcosm's applications of 2hν photon gravitational mass. Indian Journal of Physics. 2022;**96**: 1607-1611. DOI: 10.1007/s12648-021-02104-2

[29] Wesson PS. Space, Time, Matter: Modern Kaluza–Klein Theory. Singapore: World Scientific; 1999. p. 209

[30] Wesson PS. Five-Dimensional Physics. 2nd ed. Singapore: World Scientific; 2006. p. 234

[31] Overduin J, Wesson P. Kaluza-Klein gravity. Physics Reports. 1997;**283**: 303-378. DOI: 10.1016/S0370-1573(96)00046-4

[32] Overduin J, Henry RC. Physics and the pythagorean theorem. 2020. DOI: 10.48550/arXiv.2005.10671

[33] Okun LB, Selivanov KG, Telegdi VL. Gravitation, photons, clocks. Physics-Uspekhi. 1999;**42**:1045-1050. DOI: 10.1070/PU1999v042n10ABEH000597

[34] Morozov VB. On the question of the electromagnetic momentum of a charged body. Physics-Uspekhi. 2011;**54**:371-374. DOI: 10.3367/UFNe.0181.201104c.0389

[35] Tsipenyuk DY, Andreev VA. Lienard-Wiechert potentials for model of extended space. Bulletin of the Lebedev Physics Institute. 2002;**6**:1. DOI: 10.48550/arXiv.physics/0302006

[36] Haken H, Wolf HC, Brewer WD. The Physics of Atoms and Quanta. 7th ed. Berlin: Springer; 2005. p. 529. DOI: 10.1007/3-540-29281-0

[37] Dehmelt H. A single atomic particle forever floating at rest in free space: New value for electron radius. In: Workshop and Symposium on the Physics of Low-

Energy Stored and Trapped Particles (PSTP-87); 14–18 June 1987. Stockholm, Sweden: Physica Scripta; 1988. pp. 102-110. DOI: 10.1088/0031-8949/1988/T22/016

[38] Maitland A, Dunn MH. Laser Physics. Amsterdam: North-Holland Pub. Co.; 1969. p. 413

[39] Landau LD, Lifshitz EM. Mechanics. 3rd ed. Oxford: Butterworth-Heinemann; 2000. p. 20

[40] Frolov VP. Generalized Fermat's principle and action for light rays in a curved spacetime. Physical Review D. 2013;**88**:06403910. DOI: 10.1103/PhysRevD.88.064039

[41] Rivlin LA. Photons in a waveguide (some thought experiments). Physics-Uspekhi. 1997;**40**:291. DOI: 10.3367/UFNr.0167.199703g.0309

[42] Faraoni V, Dumse RM. The gravitational interaction of light: From weak to strong fields. General Relativity and Gravitation. 1999;**31**:91. DOI: 10.1023/A:1018867405133

[43] Andreev VA, Tsipenyuk DY. Mechanism of the dark matter and condensed bubble objects formationin the model of extended space. Kratk Soobshch. Fiz. 2004;**9**:13

[44] Michelson PF, Atwood WB, Ritz S. Fermi gamma-ray space telescope: High-energy results from the first year. Reports on Progress in Physics. 2010;**73**:074901. DOI: 10.1088/0034-4885/73/7/074901

[45] Cerutti B, Dubus G, Malzac J, Szostek A, Belmont R, Zdziarski A, et al. Absorption of high-energy gamma rays in cygnus X-3. Astronomy & Astrophysics. 2011;**529**:A120. DOI: 10.1051/0004-6361/201116581

[46] Abdo AA et al. Fermi gamma-ray space telescope observations of the gamma-ray outburst from 3C454.3 in November 2010. The Astrophysical Journal. 2011;**733**:L26. DOI: 10.1088/2041-8205/733/2/L26

[47] Thompson DJ, Baldinib L, Uchiyamac Y. Cosmic ray studies with the Fermi Gamma-ray Space Telescope Large Area Telescope. Astroparticle Physics. 2012;**39-40**:22-32. DOI: 10.1016/j.astropartphys.2011.10.009

[48] Misner CW, Thorne KS, Wheeler JA. Gravitation. San-Francisco: Freeman and Company; 1973. p. 1278

[49] Mazur PO, Mottola E. Surface tension and negative pressure interior of a non-singular 'black hole'. Classical and Quantum Gravity. 2015;**32**:215024. DOI: 10.1088/0264-9381/32/21/215024

[50] Tsipenyuk DY, Belyaev VB. Homogeneous Spher with excited vacuum pressure, applications in extended space model and cosmology. In: Dong S-H, editor. New Trends in Physical Science Research. Vol. 6. Hong Kong: BP International; 2022. pp. 144-155. DOI: 10.9734/bpi/ntpsr/v6/2558A

[51] Tsipenyuk DY, Belyaev VB. Dependence of universe deceleration parameter on cosmological constant: Mechanism of vacuum pressure excitation by matter. Romanian Reports in Physics. 2020;**72**:113

Chapter 3

Analytical Description of Unified Field Theory for Electromagnetic and Gravity Fields with the Introduction of Quantized Spacetime and Zero-Point Energy

Shinichi Ishiguri

Abstract

We have previously derived the quantized Einstein's gravity (QEG) equation using concepts of zero-point energy and quantized space times. The theory section in this chapter provides an analytical solution of the QEG equation that implies conservation of angular momentum in terms of quantized space times. Moreover, the temperature of the cosmic microwave background (CMB) emission is obtained, and the QEG equation solution results in an analytical (not numerical) derivation of a gravity wave. We have also analytically attempted to calculate every equation in terms of electromagnetic and gravity fields using the QEG equation solution. In the Results section of this chapter, we first confirmed that the CMB emission temperature agrees with measured values. Then, the analytical solution of the QEG equation resulted in most electromagnetic and gravity field laws, in addition to the analytically derived gravity wave, which agrees well with recent measurements. Moreover, calculations of energies in the basic configuration of quantized space times resulted in the rest energies of all three leptons. Considering this basic configuration is uniformly distributed everywhere in the universe, we can conclude that τ-particles or static magnetic field energy derived from the basic configuration of quantized space times is dark energy, which is also distributed uniformly in the universe.

Keywords: unified field theory, zero-point energy, quantized space time, quantized Einstein's gravity equation, gravity wave

1. Introduction

1.1 Content summary, including previous works

This paper introduces the concepts of quantized space times and zero-point energy. We have succeeded in reinforcing our previously established unified particle theory [1, 2] and provided the reason for three generations of leptons with these concepts. Furthermore, these concepts result in the quantization of Einstein's gravity (QEG) equation, and its analytical solution imply conservation of angular momentum in terms of quantized spacetimes. This solution solves current universe problems, such as dark energy, analytical gravity waves, etc., and creates most laws and equations regarding electromagnetic and gravity fields. Electromagnetic and gravity fields are related to weak interaction, strong interaction, neutrinos, quarks, and protons because these fields are also created from the zero-point energy [1, 2], i.e., static fields. Here, we reinforce the unified field theory introduced in the previous paper and present the basic principle that the conservation of angular momentum in terms of quantized space times, i.e., both zero-point energy and quantized space times, creates most laws regarding particle physics.

1.2 Background

In our previous papers [1, 2], we succeeded in describing most electromagnetic, gravity, weak, and strong interactions using zero-point energy and quantized space times with no numerical or fitting methods. These descriptions were found to be in agreement with measurements. In another paper [3], we analytically described neutrino self-energy and their oscillations, which also agreed with measurements.

However, in these previous papers, we did not describe the following.

1. Rotations of quantized space times using the QEG equation derived in [1].

2. Comparisons with measurements that prove the existence of the proposed quantized spacetimes.

3. Our neutrino theory [3] depends on the assumption that masses of the three leptons are given.

4. The definition of dark energy in view of particle physics.

Regarding the three lepton masses, we succeed in obtaining their values in the present chapter from the basic configuration of quantized space times. This result is important because our presented concept of quantized space times is certified by measurements. Additionally, we can conclude that the energy of this configuration of quantized space times implies dark energy because dark energy generally distributes uniformly. Furthermore, this configuration also implies that the static magnetic field energy (in GeV order) can explain recent measurements [4, 5] wherein there are static magnetic fields everywhere in the universe, even in non-macroscopic objects.

One significant point of this paper is that we succeed in obtaining the analytical (not numerical) solution of Einstein's gravity equation. The introduction of quantized space times results in the QEG equation, which enables us to analytically solve this equation. The resultant facts from this analytical solution are as follows.

1. The temperature for cosmic microwave background (CMB) emission is predicted and agrees with measurements.

2. The gravitational wave, which has previously been calculated only using numerical methods, is calculated analytically.

3. With the concept of quantized space times and the QEG equation solution, the conservation of angular momentum of quantized spacetimes creates most laws in electromagnetism and gravity. That is, the unified field theory of particle physics is now reinforced with our previous papers [1, 2].

Here, let us consider the problems in the current standard big-bang model.

1. The current model cannot explain acceleration expansion in the universe with quantity [6].

2. There is a light-element problem in the standard model. The prediction for the amount of Li (lithium) by the standard big-bang model does not agree well with recent measurements [7].

3. As mentioned earlier, CMB emissions are well described without the standard big-bang model.

4. The most serious problem with the standard big-bang model is that it must assume infinite energy in the universe considering the singularity. This assumption is strong in all general physics equations because all physics equations generally form under conservation of energy.

5. The big-bang model does not describe dark energy, which is clarified by the current study. However, this paper claims that this energy is merely a well-known particle that obeys general gravitational law. Therefore, this paper claims that dark energy, which exhibits repulsive forces, does not exist.

In short, the standard big-bang model cannot describe recent cosmology problems and is not supported by measurements. In particular, the abovementioned "Li problem" is serious. Therefore, a new model has been recently pursued by other researchers.

1.3 Summary of the significance of the present paper

We have succeeded in confirming the existence of the basic configuration of quantized spacetimes, and the concepts of quantized space times and zero-point energy have resulted in an analytical solution of Einstein's gravity equation. This implies conservation of the angular momentum of quantized space times. This solution creates most electromagnetic and gravity field laws and equations. In our previous papers [1, 2], weak interaction, strong interaction, and particle fields are well described using only the concepts of zero-point energy and quantized space times. Therefore, we now address an important principle: most physical fields and their laws are created only by conservation of angular momentum in terms of quantized space times, i.e., zero-point energy with the introduction of quantized space times.

This paper was also able to obtain the reason why leptons and neutrinos have three generations, which has been a puzzle since particle physics was established. Additionally, the main problems in cosmology have been solved here without the standard big-bang model. In particular, the gravity wave was obtained analytically.

2. Theory

2.1 Review of the concepts of quantized spacetimes and Einstein's gravity equation

2.1.1 Quantized spacetime concept

We begin with the result of the Dirac equation, which implies that a photon creates an electron and a positron:

$$\hbar\omega_0 = 2m_e c^2, \tag{1}$$

where ω_0, m_e, and c denote a constant angular frequency, the mass of an electron, and the speed of light, respectively. This equation can be interpreted as

$$\frac{1}{2}\hbar\omega_0 = m_e c^2 \tag{2}$$

and produces the minimum quantized length, λ_0, and time t_0 in terms of a space time:

$$\lambda_0 = \frac{\hbar}{2m_e c} \tag{3}$$

and

$$t_0 = \frac{\hbar}{2m_e c^2}. \tag{4}$$

We derive a more general constant quantized space-time length and time:

$$\lambda_c = \lambda_0\sqrt{1 - \frac{v^2}{c^2}} \tag{5}$$

and

$$t_c = t_0\sqrt{1 - \frac{v^2}{c^2}}. \tag{6}$$

We consistently assume that the above length (5) and time (6) are minima, thus, they cannot be divided further. As discussed later, it was found that these concepts are supported by measurement.

In Eq. (2), the left-hand side is identical for the zero-point energy in the harmonic oscillator Hamiltonian:

$$H = \left(n + \frac{1}{2}\right)\hbar\omega_0. \tag{7}$$

As every quantum field theory argues, the first term in Eq. (7) implies alternating current electromagnetism. However, the second term, called zero-point energy (neglected in quantum field theory), is more important because of direct current

(DC) electromagnetism. Note that we will report that the first term creates Maxwell's time-dependent equations in view of different approaches from quantum field theory.

Figure 1 shows a schematic of the basic configuration of quantized space times. Two quantized space times, in terms of an electric field, are rotating with velocity v. Each quantized space time, in terms of an electric field, has embedded up- and down-spin electrons (this will become important when considering the creation of a μ-particle). Note that it is necessary to distinguish these embedded electrons from real body electrons. By rotations of embedded electrons, i.e., the rotations of the two quantized space times, another quantized space time is induced in terms of a magnetic field. This magnetic field accompanies the concept of flux (defined in the central circle in **Figure 1**), that is, another quantized space time whose radius is the same as λ_c. Therefore, it is very important to distinguish the two quantized space times as:

1. a quantized space time accompanying an embedded electron in terms of an electric field and

2. a quantized space time induced in terms of a magnetic field.

As will be discussed later, the energies of the two quantized space times are commonly expressed as zero-point energies. Force F is the Lorentz force originating from the static magnetic field, and it is identified by attractive gravity forces, F, from the gravitational field. This is important because gravity and the magnetic field are unified in this scale, which results in the quantized Einstein's gravity equation. An important note is that this configuration can be described by the QEG equation solution, as will also be discussed later.

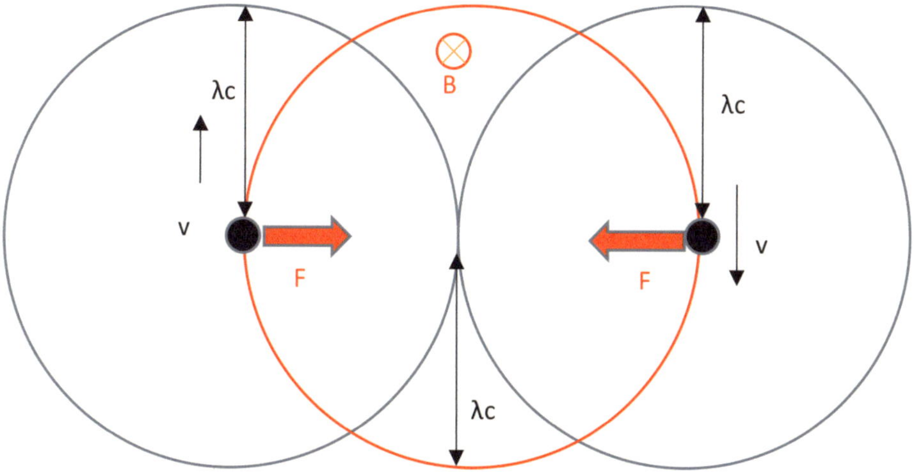

Figure 1.
Diagram of the basic configuration of quantized space times.The radius of the quantized time space is λ_c, which is determined using the Dirac equation and Lorentz contraction. For details, refer to our previous paper [1]. The black dot denotes an embedded electron, which rotates with velocity v. *The force* F *is generated as a result of magnetic field generation (the Lorentz force). Note that the magnetic flux is defined in the central circle. As discussed in our various previous papers, when the relative momentum in terms of two charged particles is zero, these two particles generally experience an attractive force that stems from the Lorentz force. As shown, this attractive force* F *is identical to an attractive gravity force, which is related to the quantization of Einstein's gravity equation.*

2.1.2 Quantization of Einstein's gravity equation

According to our previous paper [1], energy relationships in terms of gravity, static magnetic field, and electric field in the scale of the quantized space times are derived as

$$u_B = -G^2 \frac{1}{r^4} \frac{1}{c^2} u_E \left(\frac{\hbar}{2c^2} \right)^2, \; u_B = u_G \tag{8}$$

where r, G, u_B, u_G, and u_E denote the distance, the gravitational constant, magnetic field energy, gravity field energy, and electric field energy, respectively. We can combine Eq. (8) with Einstein's gravity equation because both equations include G. The existing Einstein's equation is

$$G_{\mu\nu} = \frac{8\pi G}{c^4} T_{\mu\nu} \tag{9-1}$$

and

$$G_{\mu\nu} = R_{\mu\nu} - \frac{1}{2} R g_{\mu\nu}, \tag{9-2}$$

where $R_{\mu\nu}, T_{\mu\nu}, g_{\mu\nu}$, and R denote the Riemann curvature tensor, the energy flux tensor, the metric tensor, and the Ricci tensor, respectively.

As a result of substituting G in Eq. (8), we obtain

$$G_{\mu\nu} = \frac{16\pi}{c\hbar} \sqrt{\frac{-u_B}{u_E}} \lambda_c^2 T_{\mu\nu}. \tag{10}$$

Now, it is assumed that the macroscopic tensor $g_{\mu\nu}$ is approximately the Minkowski tensor, g_{ij}, because an analytical differential cannot be defined for a quantized space time [1]. That is, it merely implies division by λ_c[1]. Similarly, the energy density is given using zero-point energy:

$$\varepsilon = \frac{1}{2} \hbar\omega / \lambda_c^3. \tag{11}$$

Thus, $T_{\mu\nu}$ is approximated using the Minkowski tensor:

$$T_{\mu\nu} = \varepsilon g_{ij}. \tag{12}$$

Considering the above, Einstein's gravitational equation is transformed to

$$R_{\mu\nu} = \left(\frac{16\pi}{c\hbar} \sqrt{\frac{-u_B}{u_E}} \frac{|u_G|}{\lambda_c} + \frac{1}{2} R \right) g_{ij}. \tag{13}$$

The energy of a quantized space time regarding a magnetic field is given as the zero-point energy:

$$|u_G| = \frac{1}{2} \hbar\omega. \tag{14}$$

Assuming the Ricci tensor, ½*R*, to be substantially smaller than the first term, we obtain

$$R_{\mu\nu} = \left(\frac{16\pi}{c\hbar}\sqrt{\frac{-u_B}{u_E}}\frac{1}{\lambda_c}\frac{1}{2}\hbar\omega\right)g_{ij}. \tag{15}$$

Moreover, considering Eq. (8), the ratio u_B/u_E in Eq. (15) is obtained:

$$u_B = -G^2\frac{1}{r^4}\frac{1}{c^2}u_E\left(\frac{\hbar}{2c^2}\right)^2 = -G^2\frac{1}{\lambda_c^4}\frac{1}{c^2}u_E\left(\frac{\hbar}{2c^2}\right)^2. \tag{16}$$

Considering this, we obtain conclusively

$$R_{\mu\nu} = \left(G\frac{8\pi}{c^4}\frac{1}{\lambda_c^3}\frac{1}{2}\hbar\omega\right)g_{ij}. \tag{17}$$

2.2 Zero-point energy in quantized spacetime for gravity or magnetic fields

Let us estimate the zero-point energy in terms of a magnetic or gravity field as well as in terms of an electric field concerning quantized space time. Here we consider energy level $|\Delta|$from special relativity:

$$|\Delta| = \frac{m_e c^2}{\sqrt{1-\frac{v^2}{c^2}}}. \tag{18}$$

This energy level is located at the middle position within the band gap of the vacuum, i.e., the energy gap is $2|\Delta|$. If $v \neq 0$ in Eq. (18), $2|\Delta|$ is produced by the product of both the fine-structure constant, α, and zero-point energy:

$$2|\Delta| = \frac{1}{2}\hbar\omega \times \alpha. \tag{19}$$

This fact will be proved while discussing CMB here or can be referenced in the literature [8]. Note that if $v = 0$ in Eq. (18), this equation for $|\Delta|$ implies the zero-point energy in terms of an electric field, which is related to Eq. (2).

Conversely, v in Eq. (18) is assumed to be the critical velocity, v_c, for an electron. That is, an electron has a maximum velocity v_c less than the speed of light c when largely accelerated. According to our previous papers [1, 3], an electron can accompany an e-neutrino and, thus, the e-neutrino speed is equal to the critical velocity of an electron. Therefore, $v = v_c$ is substituted in Eq. (18) by $0.994c$ [1, 3]. Using Eq. (19), the calculated zero-point energy for the magnetic or gravity field then becomes

$$\frac{1}{2}\hbar\omega = \frac{2}{\alpha}|\Delta| = 1.23 \times 10^9 \text{ eV}. \tag{20}$$

This value agrees with the measurement of a τ-particle [9].

In summary, the zero-point energy is related to special relativity energy, Eq. (18), where $v = 0$ implies the rest energy of an electron and is related to the quantized space time in terms of an electric field. While v in Eq. (18) is the v_c for an electron, the zero-point energy in Eq. (19) implies quantized space time in terms of magnetic or gravity fields.

2.3 Three generations of lepton

2.3.1 Collapse of the basic configuration of quantized spacetimes

Figure 2 schematically indicates how the magnetic or gravity field in a quantized spacetime (i.e., the combination of two embedded electrons in two quantized space–times) collapses. Each quantized space time in terms of magnetic or gravity fields has a torque property whose moment corresponds to the magnetic field vector.

From torque properties, a larger magnetic field is generated if two magnetic field vectors of two quantized space times in terms of a magnetic field take the same direction. Generally, this maximum superposition occurs in the location of the universe at which the gravity field becomes extremely strong. On the contrary, however, if two magnetic field vectors of two quantized space times in terms of the magnetic field take reverse directions, the net magnetic field vanishes. The magnetic field energy is converted to τ-particles while μ-particle energy comes from the spin interaction of two electrons embedded in quantized space times in terms of the electric field. In this way, a quantized space time in terms of a magnetic or gravity field collapses even though the combination energy of two embedded electrons in two quantized space times is quite large [3]. This fact results in the creation of τ- and μ-particles, as discussed later.

2.3.2 Masses of μ- and τ-particles from the basic configuration of quantized spacetimes

This paper claims that the masses of the three generations of leptons stem from the abovementioned collapse of the basic configuration of quantized space times. As a result of the collapse, three energies are generated from collapsed quantized space times. Based on **Figure 1**, we claim the following points.

1. The combination energy between two embedded electrons in quantized space times in terms of electric field, i.e., the magnetic field (gravity field) energy in a quantized space time, is converted. This energy corresponds to the remaining energy of the τ-particle.

2. Each embedded electron in two quantized space times, which take rotations and induce the magnetic field energy in quantized space time, have interactions in

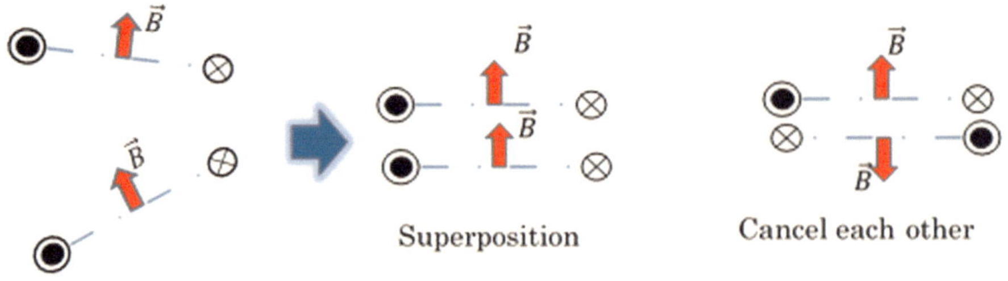

Figure 2.
Schematic wherein two quantized space times in terms of a magnetic or gravity field interact with each other and how these quantized space times collapse. This figure was cited from [10]. First, a quantized space time in terms of magnetic field has a torque property whose moment corresponds to its magnetic field vector. The superposition case in this figure is such that the two magnetic field vectors are maximally strengthened. An important case is their cancelation with each other wherein the quantized space times in terms of a magnetic field collapse.

terms of spins (up and down). This interaction is converted to the remaining energy of the μ-particle.

3. Embedded electrons in **Figure 1** automatically gain real bodies.

In the Results section, actual calculations illustrating these points will be conducted.

2.4 Analytical solution to the quantized Einstein's gravity equation

We now consider the quantized Einstein's gravity (QEG) equation.

$$R_{ij} = \frac{8\pi G}{c^4} \frac{\frac{1}{2}\hbar\omega}{\lambda_c^3} g_{ij}, \tag{21}$$

where the Minkowski tensor, g_{ij}, is given by

$$g_{ij} = \begin{pmatrix} -1 & & & 0 \\ & -1 & & \\ & & -1 & \\ 0 & & & +1 \end{pmatrix} \tag{22}$$

This QEG equation requires a specific form of the Riemann curvature tensor, R_{ij}, for the following reasons.

1. Because g_{ij} is a diagonal matrix, R_{ij} is a diagonal matrix considering Eq. (21).

2. The QEG equation must automatically express Lorentz conservation and does not include this conservation as a condition.

3. R_{ij} is a covariance tensor. Thus, it must be composed by the direct product of position vector k:

$$k = xe^1 + ye^2 + ze^3 + icte^4, \tag{23}$$

where i in the fourth term denotes the imaginary unit.

$$k \times k \to R_{ij}, \tag{24}$$

where the symbol $\times$ implies the direct product of the vectors in this paper. Since

$$e^i \bullet e^j = \delta^{ij}, \tag{25}$$

$$R_{ij} = \begin{pmatrix} x^2 & & & 0 \\ & y^2 & & \\ & & z^2 & \\ 0 & & & -(ct)^2 \end{pmatrix} \tag{26}$$

In the QEG equation, Eq. (21) takes the trace, Tr, to form Lorentz conservation:

$$x^2 + y^2 + z^2 - (ct)^2 = (-2)\frac{8\pi G}{c^4}\frac{\frac{1}{2}\hbar\omega_i}{\lambda_c^3}. \tag{27}$$

According to Eq. (27), Lorentz conservation is automatically presented. In the derived equation, time t should be considered a period,

$$r_i^2 \equiv x^2 + y^2 = (-2)\frac{8\pi G}{c^4}\frac{\frac{1}{2}\hbar\omega_i}{\lambda_c^3} + \left(c\frac{2\pi}{\omega}\right)^2 - z^2, \tag{28}$$

where r_i implies the radius of rotation at each location indexed by i, and we assume cylindrical coordinates. Note that r_i depends on the values of variable z. Moreover, variable angular frequency ω was introduced because time t implies a period. As will be described later, the derived equation implies that z gives an anisotropic property. We will see that this solution of the QEG equation describes both electromagnetism and gravity fields well.

Considering Eqs. (21), (22), and (26), each position and time variable, x, y, z, and t, are not independently defined. Thus, the Eq. (21) obtains its meaning only when the trace of both sides is considered. This fact is important because we claim that moving (rotating) space time exists. That is, the mathematical metrics in terms of Cartesian geometry are an approximated concept in a vacuum. This can be understood by considering an analogy that, while a rigid body has metrics on it like a length of a line, the area of a square, etc., at the microscopic scale of the rigid body, many thermally fluctuated lattices (phonons) exist and imply that the mathematical metrics on the rigid body are not formed at the microscopic scale.

2.5 Cosmic microwave background (CMB)

First, let us consider the analytical solution of the QEG equation again:

$$r_i^2 \equiv x^2 + y^2 = (-2)\frac{8\pi G}{c^4}\frac{\frac{1}{2}\hbar\omega_i}{\lambda_c^3} + \left(c\frac{2\pi}{\omega}\right)^2 - z^2. \tag{29}$$

Index i is associated with both r_i and $\frac{1}{2}\hbar\omega_i$. Note that here we assume $z = 0$, resulting in

$$r_i^2 \equiv (-2)\frac{8\pi G}{c^4}\frac{\frac{1}{2}\hbar\omega_i}{\lambda_c^3} + \left(c\frac{2\pi}{\omega}\right)^2. \tag{30}$$

Because r_i should be considered a macroscopic variable of radius, the first term of the right-hand side (i.e., the zero-point energy having index i) should be neglected. As a result, considering the area of a circle, the macroscopic variable radius r appears and thus a unique angular frequency is derived as

$$\omega = \frac{\sqrt{4\pi}c}{r}. \tag{31}$$

Next, the angular frequency ω in Eq. (31) are substituted into the Prank emission, α_T, Eq. (32):

$$\alpha_T = \frac{1}{\pi^2} \frac{\hbar\omega/c^2}{exp\left(\frac{\hbar\omega}{k_B T}\right) - 1}, \tag{32}$$

where T and k_B denote the temperature and the Boltzmann constant, respectively. When the exponential function in Eq. (32) becomes e^{-1}, the following equation holds:

$$\frac{k_B T_0 r}{\hbar\sqrt{4\pi c}} = 1. \tag{33}$$

In Eq. (33), temperature T_0 implies one Prank emission. We claim that a CMB photon is derived from the energy gap, which fluctuates in the energy level of the vacuum [3] and is related to the e-neutrino self-energy. That is, considering that r in Eq. (33) is the wavelength of a photon, this wavelength can be derived from fluctuations in the vacuum energy level.

Figure 3 shows the iteration wherein photons are absorbed or emitted to or from the energy gap, respectively. This implies that CMB photons are created and absorbed everywhere in the universe, and thus, we claim that CMB photons are not the source of birth of our universe in terms of the big-bang.

In the Results section, the actual calculation of the phenomena discussed above will be conducted using e-neutrino self-energy.

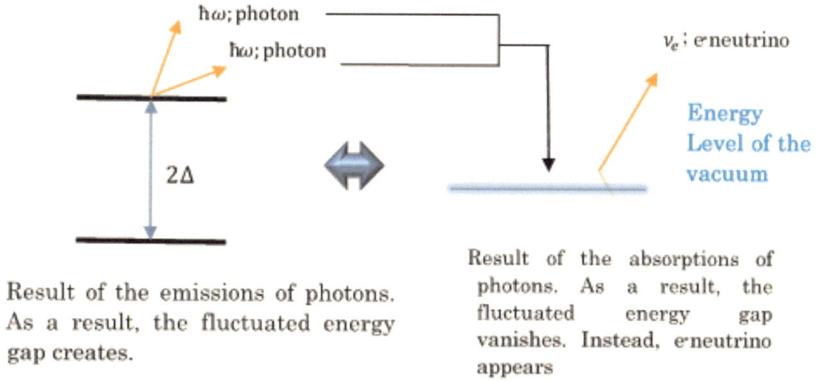

Figure 3.
Schematic representation of creation and absorption of CMB photons through the neutrino energy gap. This figure was cited from [10]. The left-hand-side panel indicates the creation of the energy gap, which fluctuates in the vacuum energy level. An energy gap is created by emission of photons from the vacuum energy level. This created energy gap is essentially equal to the self-energy of an e-neutrino. The right-hand-side panel shows the disappearance of the energy gap by absorption of photons; a real body of an e-neutrino is emitted. As mentioned in our previous paper [3], an energy gap is again created at the energy level of the vacuum according to the BCS ground state because of many-body interactions of the basic configuration of quantized space times. The left-hand and right-hand panels occur everywhere in the universe locally with iterations. Thus, the creation and absorption of photons in terms of CMB arise everywhere and the CMB source is not the birth of the universe (i.e., the big-bang).

2.6 Analytical derivation of the gravity wave

In the QEG equation solution,

$$r_i^2 = x^2 + y^2 = \frac{-16\pi G}{c^4}\frac{\frac{1}{2}\hbar\omega_i}{\lambda_c^3} + \left(c\frac{2\pi}{\omega}\right)^2 - z^2, \tag{34}$$

$r_i = 0$ is a condition of the generation of gravity waves. Note that index i is arbitrary. The condition of the generation of gravity waves implies that previous rotations cease everywhere. That is, macroscopic rotation ceased [11, 12].

Depending on z, variable angular frequency ω is varied. The center of the previous rotation is considered in this case and in cylindrical coordinates. That is, $z = 0$ is assumed. This also implies that the maximum is considered. For a secondary condition, the zero-point energy is converted to a photon by the product of the fine-structure constant, α, because it is needed to convert the energy level to an energy gap. When the uncertainty relation is introduced, an ω equation dependent on Δt is derived:

$$\frac{1}{2}\hbar\omega_i \rightarrow \frac{1}{2}\hbar\omega_i\alpha = n\hbar\omega = \hbar(n\omega) = |\Delta| \tag{35}$$

and

$$|\Delta| \times \Delta t \approx \hbar. \tag{36}$$

Considering Eq. (35), $\omega \rightarrow n\omega$ and

$$\left(c\frac{2\pi}{n\omega}\right)^2 = \frac{16\pi G}{c^4}\frac{1}{\lambda_c^3}|\Delta| \tag{37}$$

and

$$\left(c\frac{2\pi}{n\omega}\right)^2 \Delta t = \frac{16\pi G}{c^4}\frac{1}{\lambda_c^3}\hbar. \tag{38}$$

In this equation, the distributed relationship regarding λ_c is employed:

$$(n\omega)^5 = \frac{\pi c^9}{4G}\frac{1}{\hbar}\Delta t. \tag{39}$$

Moreover, quantum number n is defined as

$$\sum_n t_0 \equiv 1[s], \tag{40-1}$$

That is,

$$nt_0 \equiv 1. \tag{40-2}$$

Next, strain h_{max} is considered:

$$h_{max} = \frac{\delta L}{\frac{L}{2}}. \tag{41}$$

This definition is translated to

$$h_{max} = \frac{\lambda_c}{L/2}, \tag{42}$$

where

$$\lambda_c = ct_c, \tag{43-1}$$

$$t_c = t_0\sqrt{1 - \frac{v^2}{c^2}}, \tag{43-2}$$

and

$$v = v_c = 0.994c. \tag{43-3}$$

As mentioned in Section 2.2, v_c is the critical speed of an electron and is equal to that of an e-neutrino. Therefore, this expression implies consideration of quantized space times in terms of gravity (magnetic) fields.

With

$$\frac{L}{2} = c\Delta t \tag{44}$$

we have

$$h_{max} = \frac{t_c}{\Delta t} \tag{45}$$

and the chirp signal is given by

$$u_p = \frac{t_c}{\Delta t} cos\,(\omega t_{00}), \tag{46}$$

where t_{00} is defined as the constant $1[s]$ because ω is a variable dependent on Δt. The chirp signal will be further discussed in the Results section.

2.7 Unified field picture in terms of electromagnetic and gravity fields by rotations of quantized spacetimes

The solution of the QEG equation is again given as

$$r_i^2 = x^2 + y^2 = \left(c\frac{2\pi}{\omega}\right)^2 - \frac{16\pi G}{c^4}\frac{1}{\lambda_c^3}\frac{1}{2}\hbar\omega_i, \tag{47}$$

where $z = 0$ is assumed.

As mentioned earlier, this equation implies quantized time-space rotation.

2.7.1 The case of direct current

2.7.1.1 General notation

Eq. (47) creates every equation regarding electromagnetism and Newtonian gravity. To show this, λ_c should first be canceled and later we consider general fields. We assume that electric and magnetic fields only in a quantized space time have energies

$$\frac{1}{2}\hbar\omega_i = \frac{1}{2}\varepsilon_0 E_i^2 \lambda_c^3 \tag{48-1}$$

and

$$\frac{1}{2}\hbar\omega_i = \frac{B_i^2}{2\mu_0}\lambda_c^3, \tag{48-2}$$

respectively. Concerning gravity, in the Section 2.2, we derived the zero-point energy in terms of the gravity field:

$$\frac{1}{2}\hbar\omega_i = \frac{2}{\alpha}|\Delta| = \frac{2}{\alpha}\frac{m_e c^2}{\sqrt{1-\frac{v^2}{c^2}}}. \tag{48-3}$$

Furthermore, the general wave function is considered:

$$\int |\psi|^2 dv = |\psi_i|^2 \lambda_c^3 = 1. \tag{48-4}$$

Note that the differential and integral become merely division and product, respectively, in the quantized space time [1].

Each of the above equations is substituted into Eq. (47) and the general electric, magnetic, and gravity field equations are derived as follows:

$$r_i^2 = \left(c\frac{2\pi}{\omega}\right)^2 - \frac{16\pi G}{c^4}\frac{1}{2}\varepsilon_0 E_i^2, \tag{49-1}$$

$$r_i^2 = \left(c\frac{2\pi}{\omega}\right)^2 - \frac{16\pi G}{c^4}\frac{B_i^2}{2\mu_0}, \tag{49-2}$$

and

$$r_i^2 = \left(c\frac{2\pi}{\omega}\right)^2 - \frac{16\pi G}{c^4}|\psi_i|^2\frac{2}{\alpha}\frac{m_e c^2}{\sqrt{1-\frac{v^2}{c^2}}}, \tag{49-3}$$

respectively.

2.7.1.2 Derivation of each Poisson equation

The Poisson equations can be obtained in terms of electrostatic, vector, and gravity potentials based on the results obtained in the previous section.

First, consider the Poisson equation in terms of electrostatic potential:

$$r_i^2 = \left(c\frac{2\pi}{\omega}\right)^2 - \frac{16\pi G}{c^4}\frac{1}{2}\varepsilon_0 E_i^2. \tag{49-1}$$

If the first term is neglected,

$$r_i \equiv \lambda_c. \tag{50}$$

Thus,

$$\frac{1}{2}\varepsilon_0\frac{E_i^2}{\lambda_c^2} = \frac{-c^4}{16\pi G}. \tag{51-1}$$

In Eq. (51-1), division by λ_c must be translated to the normal differential to express the mathematic equation. That is, the differential is revived [1]:

$$\frac{1}{2}\varepsilon_0\frac{dE_i^2}{dr^2} = \frac{-c^4}{16\pi G}. \tag{51-2}$$

Thus,

$$\varepsilon_0 E_r\frac{dE_r}{dr}\frac{1}{\lambda_c} = \frac{-c^4}{16\pi G}. \tag{52}$$

Considering the concept of quantized space times:

$$r = n\lambda_c, \tag{53}$$

and

$$nE_{r0}\frac{dE_r}{dr}\frac{1}{r} = \frac{-1}{\varepsilon_0}\frac{c^4}{16\pi G}. \tag{54}$$

Introducing the electrostatic potential Φ, we have

$$E_r = \frac{-d\Phi}{dr} \tag{55}$$

and

$$\frac{-d^2\Phi}{dr^2} = \frac{-1}{\varepsilon_0}\frac{c^4}{16\pi G}\frac{1}{nE_{r0}}r. \tag{56}$$

Herein, the following relation is assumed:

$$E_{r0} = -vB_{z0}, \tag{57}$$

where v implies an arbitrary and rotational velocity, not the speed of light c, and Eq. (56) becomes

$$\frac{d^2\Phi}{dr^2} = \frac{-1}{\varepsilon_0}\frac{c^4}{16\pi G}\frac{1}{v}\frac{1}{B_{z0}}\frac{1}{n}r. \tag{58}$$

Considering the cyclotron angular frequency,

$$\omega_c = \frac{eB_{z0}}{m_e}, \tag{59}$$

we have

$$\frac{d^2\Phi}{dr^2} = \frac{-1}{\varepsilon_0}\frac{c^4}{16\pi G}\frac{1}{v}\frac{e}{m_e\omega_c}\frac{1}{n}r. \tag{60}$$

Or, using

$$v = r\omega_c, \tag{61}$$

we obtain

$$\frac{d^2\Phi}{dr^2} = \frac{-1}{\varepsilon_0}\frac{c^4}{16\pi G}\frac{e}{m_e\omega_c^2}\frac{1}{n}. \tag{62}$$

Next, we consider the Poisson equation in terms of vector potential. Similar to the case of an electric field and using Eq. (49-2), we obtain

$$B_z\frac{dB_z}{dr^2} = -\mu_0\frac{c^4}{16\pi G} \tag{63}$$

and

$$B_z\frac{dB_z}{dr}\frac{1}{\lambda_c} = nB_{z0}\frac{dB_z}{dr}\frac{1}{r} = -\mu_0\frac{c^4}{16\pi G}. \tag{64}$$

Cylindrical coordinates are considered in this case and, thus, the component of a vector potential is introduced by

$$B_z = \frac{1}{r}A_\varphi. \tag{65}$$

Calculating $\frac{dB_z}{dr}$ we obtain

$$nB_{z0}\left(\frac{-A_\varphi}{r^2}\right) = -\mu_0\frac{c^4}{16\pi G}r. \tag{66}$$

Thus, A_φ becomes a special dependent and division of r is transformed into the differential:

$$\frac{d^2A_\varphi}{dr^2} = \mu_0\frac{c^4}{16\pi G}\frac{1}{nB_{z0}}r. \tag{67}$$

With the introduction of cyclotron angular frequency ω_c we have

$$\frac{d^2 A_\varphi}{dr^2} = \mu_0 \frac{c^4}{16\pi G} \frac{1}{n} \frac{e}{m_e \omega_c} r. \tag{68}$$

Consequently,

$$\frac{d^2 A_\varphi}{dr^2} = \mu_0 \frac{c^4}{16\pi G} \frac{1}{n} \frac{e}{m_e \omega_c} r. \tag{69}$$

Next, we derive the Poisson equation in terms of gravity beginning with

$$r_i^2 = \left(c\frac{2\pi}{\omega}\right)^2 - \frac{16\pi G}{c^4} |\psi_i|^2 \frac{2}{\alpha} \frac{m_e c^2}{\sqrt{1 - \frac{v^2}{c^2}}}. \tag{49-3}$$

The first term is neglected and $r_i \equiv \lambda_c$ is assumed, resulting in

$$\lambda_c^2 = \frac{-16\pi G}{c^4} |\psi_i|^2 \frac{2}{\alpha} \frac{m_e c^2}{\sqrt{1 - \frac{v^2}{c^2}}}. \tag{70}$$

In short,

$$1 = \frac{-16\pi G}{c^4} \frac{|\psi_i|^2}{\lambda_c^2} \frac{2}{\alpha} \frac{m_e c^2}{\sqrt{1 - \frac{v^2}{c^2}}}. \tag{71-1}$$

Then, the normal differential must be revived [1] to ensure the mathematical expression:

$$1 = \frac{-16\pi G}{c^4} \frac{d^2 |\psi_i|^2}{dr^2} \frac{2}{\alpha} \frac{m_e c^2}{\sqrt{1 - \frac{v^2}{c^2}}}. \tag{71-2}$$

The following term for the potential energy for gravity, Φ_G,

$$g\delta\left(\vec{r}\right)|\psi_i|^2 \equiv \Phi_G, \tag{72}$$

where g is a variable and δ is the Dirac function, is introduced.
Thus,

$$\frac{d^2 \Phi_G}{dr^2} = \frac{-c^4}{16\pi G} \frac{\alpha}{2} \frac{\sqrt{1 - \frac{v_R^2}{c^2}}}{m_e c^2} g\delta\left(\vec{r}\right), \tag{73}$$

where $v \equiv v_R$ denotes the relative velocity between two charged particles because $\boldsymbol{r}$ in the Dirac function implies the relative distance between the two charged particles.

When relative velocity v_R is zero, two charged particles generally experience a strong attractive force with each other as the Lorentz force. For example, this attractive force creates a Cooper pair in high-temperature superconductors [13]. Therefore, when v_R is assumed to be zero, Eq. (73) is approximated as

$$\frac{d^2\Phi_G}{dr^2} = \frac{-1}{16\pi G}\frac{1}{m_e}c^2 g\delta\left(\vec{r}\right). \tag{74}$$

In the above equation, g should be considered a variable. Note that $\frac{\alpha}{2}$ has meaning only when v_R is not zero but large; the above conclusive equation does not include this fine-structure constant.

Moreover,

$$\int \delta\left(\vec{r}\right)dv = 1. \tag{75}$$

Considering Eq. (75), we take the volume integral to Eq. (74). Note that, spherical coordinates are considered in this case because r implies a relative distance. We obtain

$$\int \frac{d^2\Phi_G}{dr^2} 4\pi r^2 dr = \frac{-1}{16\pi G}\frac{1}{m_e}c^2 g \tag{76}$$

with the left side equal to

$$\int d\Phi_G \frac{d}{dr} 4\pi r^2 = 8\pi r \Phi_G. \tag{77}$$

Thus, we finally obtain a Newtonian equation:

$$\Phi_G = \frac{-1}{8\pi}\frac{1}{16\pi G}\frac{c^2}{m_e} g\frac{1}{r}. \tag{78}$$

In the Results section, we will examine the validity of these derived Poisson equations using actual calculations.

2.7.2 Derivation in the case of alternating current

First, when the zero-point energy in the QEG equation solution is translated to a photon, the energy gap is expressed as

$$\frac{1}{2}\hbar\omega_i \rightarrow \frac{1}{2}\hbar\omega_i\alpha = |\Delta| = |E_i - E_j| \tag{79}$$

and the basic solution becomes

$$r_i^2 = \left(c\frac{2\pi}{\omega}\right)^2 - \frac{16\pi G}{c^4}\frac{1}{\lambda_c^3}|E_i - E_j|, \tag{80}$$

where

$$|E_i - E_j| = \int \left| \frac{1}{2}\varepsilon_0 E_i^2 - \frac{B_j^2}{2\mu_0} \right| dv = \left| \frac{1}{2}\varepsilon_0 E_i^2 - \frac{B_j^2}{2\mu_0} \right| \lambda_c^3. \tag{81}$$

Thus,

$$r_i^2 = \left(c\frac{2\pi}{\omega} \right)^2 - \frac{16\pi G}{c^4} \left| \frac{1}{2}\varepsilon_0 E_i^2 - \frac{B_j^2}{2\mu_0} \right|. \tag{82}$$

As shown in **Figure 4**, Eq. (82) implies that the magnetic field energy indexed by j is dependent on the electric field energy indexed by i (here we do not consider the previously mentioned and basic configuration of quantized spacetimes). Thus, a magnetic field is induced by an electric field. If indices i and j are altered, the electric field energy becomes dependent and is induced from the magnetic field energy:

$$r_i^2 = \left(c\frac{2\pi}{\omega} \right)^2 - \frac{16\pi G}{c^4} \left| \frac{1}{2}\varepsilon_0 E_i^2 - \frac{B_j^2}{2\mu_0} \right| \text{ or } r_j^2 = \left(c\frac{2\pi}{\omega} \right)^2 - \frac{16\pi G}{c^4} \left| \frac{-1}{2}\varepsilon_0 E_j^2 + \frac{B_i^2}{2\mu_0} \right|. \tag{83}$$

That is, two energy levels indexed by both i and j are induced by each other and are iterated by ω. This physical picture describes the process of an electromagnetic wave and Eq. (83) implies time-dependent Maxwell's equations.

Now we create Maxwell's time-dependent equations based on Eq. (82):

$$\omega \to \frac{2\pi}{t_c}, \tag{84}$$

$$r_i \to \lambda_c, \tag{85}$$

and

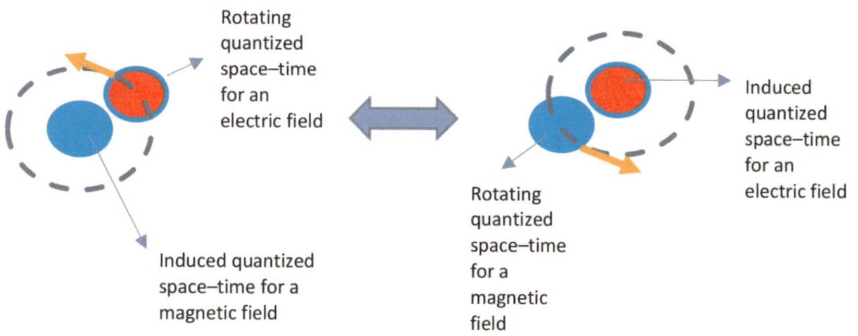

Figure 4.
Schematic of induction of quantized space times in terms of both electric and magnetic fields. This figure was cited from [10]. The left panel indicates that rotation of quantized space time for an electric field induces a quantized space time for a magnetic field. On the contrary, the right panel shows that rotation of the quantized space time for a magnetic field induces a quantized space time for an electric field.As will be derived later, these phenomena imply the time-dependent Maxwell's equations and indicate the process of electromagnetic wave induction.

$$\lambda_c^2 = (ct_c)^2 - \frac{16\pi G}{c^4}\left(\frac{1}{2}\varepsilon_0 E^2 - \frac{B^2}{2\mu_0}\right). \tag{86}$$

In Eq. (86), E and B are not magnitudes but components. Considering generalizations into three dimensions in view of vector analysis, these components are assumed to be arbitrary regarding any coordinates. Eq. (86) can be written as

$$\lambda_c^2 - \frac{16\pi G}{c^4}\left(\frac{B^2}{2\mu_0}\right) = (ct_c)^2 - \frac{16\pi G}{c^4}\left(\frac{1}{2}\varepsilon_0 E^2\right) \tag{87}$$

and from this equation we consider the following simultaneous equations:

$$\lambda_c^2 - \frac{16\pi G}{c^4}\left(\frac{B^2}{2\mu_0}\right) = \alpha. \tag{88-1}$$

and

$$(ct_c)^2 - \frac{16\pi G}{c^4}\left(\frac{1}{2}\varepsilon_0 E^2\right) = \alpha. \tag{88-2}$$

Eq. (88-1) becomes

$$1 - \frac{16\pi G}{c^4}\frac{1}{2\mu_0}\frac{B^2}{\lambda_c^2} = \frac{\alpha}{\lambda_c^2}. \tag{89}$$

The differential must be revived and the number one is ignored to obtain:

$$\frac{-16\pi G}{c^4}\frac{1}{2\mu_0}\frac{dB^2}{dr^2} = \frac{d\alpha}{dr^2}, \tag{90}$$

$$\frac{-16\pi G}{c^4}\frac{1}{2\mu_0}2B\frac{dB}{dr}\frac{1}{\lambda_c} = \frac{d\alpha}{dr}\frac{1}{\lambda_c}, \tag{91}$$

and

$$\frac{-16\pi G}{c^4}\frac{1}{\mu_0}\frac{dB}{dr}B = \frac{d\alpha}{dr}. \tag{92}$$

Eq. (88-2) becomes

$$1 - \frac{16\pi G}{c^4}\frac{1}{2}\varepsilon_0\frac{E_i^2}{ct_c^2} = \frac{\alpha}{ct_c^2}, \tag{93}$$

and

$$\frac{-16\pi G}{c^4}\frac{1}{2}\varepsilon_0\frac{1}{c}\frac{dE^2}{dt}\frac{1}{t_c} = \frac{d\alpha}{dt}\frac{1}{c}\frac{1}{t_c} \tag{94}$$

and

$$\frac{-16\pi G}{c^4}\varepsilon_0 E\frac{dE}{dt}=\frac{d\alpha}{dt}. \tag{95}$$

At this time, the following Lorentz conservation is assumed:

$$(dr)^2-c^2(dt)^2\equiv 0. \tag{96-1}$$

That is,

$$dr=\pm c dt. \tag{96-2}$$

The sign + is employed and Eq. (95) becomes

$$\frac{-16\pi G}{c^4}\varepsilon_0 E\frac{dE}{dt}=c\frac{d\alpha}{dr}. \tag{95-2}$$

Combining the above with Eq. (92):

$$\frac{-16\pi G}{c^4}\varepsilon_0 E\frac{dE}{dt}=c\left(\frac{-16\pi G}{c^4}\frac{1}{\mu_0}B\frac{dB}{dr}\right) \tag{97}$$

and

$$\varepsilon_0 E\frac{dE}{dt}=c\frac{1}{\mu_0}B\frac{dB}{dr}. \tag{98}$$

The ratio E/B is related to the characteristic impedance Z in the vacuum and is calculated as

$$\frac{E}{B}=\frac{E}{\mu_0 H}=\frac{1}{\mu_0}Z=\frac{1}{\mu_0}\sqrt{\frac{\mu_0}{\varepsilon_0}}=\frac{1}{\sqrt{\mu_0\varepsilon_0}}=c. \tag{99}$$

Considering this relation, Eq. (98) becomes

$$\frac{dE}{dt}=\frac{1}{\varepsilon_0\mu_0}\frac{dB}{dr}. \tag{100}$$

In view of vector analysis, this process can be generalized into three dimensions as

$$rot\vec{H}=\frac{\partial\vec{D}}{\partial t}. \tag{101}$$

This conclusive equation is identical to Maxwell's third equation.
Now we obtain Maxwell's forth equation using the same method.
In the QEG equation solution, indices i and j are altered such that

$$r_j^2=\left(c\frac{2\pi}{\omega}\right)^2-\frac{16\pi G}{c^4}\left(\frac{-1}{2}\varepsilon_0 E_j^2+\frac{B_i^2}{2\mu_0}\right). \tag{102}$$

In a similar process, the following simultaneous equations are formed:

$$\lambda_c^2 - \frac{16\pi G}{c^4}\left(\frac{1}{2}\varepsilon_0 E_j^2\right) = \alpha \tag{103-1}$$

and

$$(ct_c)^2 - \frac{16\pi G}{c^4}\left(\frac{B_i^2}{2\mu_0}\right) = \alpha. \tag{103-2}$$

From Eq. (103-1) we have

$$1 - \frac{16\pi G}{c^4}\frac{1}{2}\varepsilon_0 \frac{E^2}{\lambda_c^2} = \frac{\alpha}{\lambda_c^2}. \tag{104}$$

As mentioned earlier, E and B are arbitrary components of vectors regardless of any coordinate system (not magnitude).

From the division of λ_c, the differential must be revived and the number one is ignored to obtain:

$$\frac{-16\pi G}{c^4}\frac{1}{2}\varepsilon_0 \frac{dE^2}{dr^2} = \frac{d\alpha}{dr^2}, \tag{105}$$

$$\frac{-16\pi G}{c^4}\varepsilon_0 E\frac{dE}{dr}\frac{1}{\lambda_c} - \frac{d\alpha}{dr}\frac{1}{\lambda_c}, \tag{106}$$

and

$$\frac{-16\pi G}{c^4}\varepsilon_0 E\frac{dE}{dr} = \frac{d\alpha}{dr}. \tag{107}$$

Eq. (103-2) becomes

$$1 - \frac{16\pi G}{c^4}\frac{1}{2\mu_0}\frac{B^2}{ct_c^2} = \frac{\alpha}{ct_c^2} \tag{108}$$

and, similarly,

$$\frac{-16\pi G}{c^4}\frac{1}{2\mu_0}\frac{dB^2}{cdt^2} = \frac{d\alpha}{cdt^2}, \tag{109}$$

$$\frac{-16\pi G}{c^4}\frac{1}{\mu_0}B\frac{dB}{dt}\frac{1}{c}\frac{1}{t_c} = \frac{d\alpha}{dt}\frac{1}{t_c}\frac{1}{c}, \tag{110}$$

and

$$\frac{-16\pi G}{c^4}\frac{1}{\mu_0}B\frac{dB}{dt} = \frac{d\alpha}{dt}. \tag{111}$$

From the abovementioned Lorentz conservation,

$$dr = \pm cdt. \tag{96-2}$$

In this case, the sign – is employed and Eq. (111) becomes

$$\frac{-16\pi G}{c^4}\frac{B}{\mu_0}\frac{dB}{dt} = -c\frac{d\alpha}{dr}. \tag{112}$$

Combining the above equation with Eq. (107),

$$\frac{-16\pi G}{c^4}\frac{B}{\mu_0}\frac{dB}{dt} = -c\left(\frac{-16\pi G}{c^4}\varepsilon_0 E\frac{dE}{dr}\right) \tag{113}$$

and

$$\frac{B}{\mu_0}\frac{dB}{dt} = -c\varepsilon_0 E\frac{dE}{dr}. \tag{114}$$

As mentioned earlier,

$$\frac{E}{B} = c, \tag{99}$$

thus,

$$\frac{dB}{dt} = -c^2\mu_0\varepsilon_0\frac{dE}{dr} = \frac{-dE}{dr}. \tag{115}$$

In view of vector analysis, this equation can be generalized to three dimensions as:

$$\frac{\partial\vec{B}}{\partial t} = -rot\vec{E}. \tag{116}$$

This is how we derive Maxwell's forth equation.
In the Results section, we will summarize these processes and results.

3. Results

3.1 Masses of the three leptons

From our previous paper [3], the combination energy (i.e., Lorentz force) in terms of two embedded electrons in quantized space time, i.e., the magnetic (gravity) field energy U_B is estimated as.

$$|U_B| = |-8.0 \times 10^{-10}|\,\mathrm{J} \tag{117}$$

This energy gives the rest energy of τ-particles. Considering that τ-particles are fermions,

$$|U_B| = 2M_\tau c^2, \tag{118}$$

where M_τ is the mass of a τ-particle.

As a result,

$$M_\tau = 2.5 \times 10^9 \text{eV}/\text{c}^2. \tag{119}$$

Comparing the above result with the measurement in [9] indicates that the theoretical value has the same order as the measurement value but is slightly larger. This is because the gravity interaction between two τ-particles in the theoretical value is due to their large masses. Strictly speaking, a very small term regarding gravity interaction between two τ-particles should be added to Eq. (118). As mentioned, we also claim that there are attractive, not repulsive, dark energy interactions due to gravity.

We next consider the case of μ-particles. From our previous paper regarding superconductivity [14], the spin interaction, V, between up- and down-spin electrons is expressed as

$$V = \frac{-e^2\hbar^2}{16\pi^2 {m_e}^2 (2\lambda_c)^3}, \tag{120}$$

where

$$\lambda_c \approx \lambda_0, \tag{121}$$

and

e denotes the electron charge. Considering that the μ-particle is a fermion and that, from **Figure 1**, the relative distance in Eq. (120) is the diameter of quantized space–time $2\lambda_c$, the remaining energy is derived as

$$2M_\mu c^2 = \left| \frac{-e^2\hbar^2}{16\pi^2 m_e^2 (2\lambda_c)^3} \right|, \tag{122}$$

where M_μ denotes the mass of a μ-particle:

$$M_\mu = 1.15 \times 10^8 eV/c^2. \tag{123}$$

The above value is in sufficient agreement with the measurement provided in [9]. Note that a real electron appears automatically as a result of the collapse of the configuration of quantized space times.

The significance of the above discussion is that we have clarified the reason why leptons have three generations from the view of the basic configuration of quantized spacetimes (**Figure 1**). In a previous paper [3], we calculated the self-energy of three-generation neutrinos. Thus, with these results, we have now obtained a comprehensive understanding of why leptons have three generations.

3.2 CMB emission

The theory section derived the following unique angular frequency:

$$\omega = \frac{\sqrt{4\pi c}}{r}. \tag{31}$$

Thus, when the exponential function in Eq. (32) becomes e^{-1}, the following equation is obtained:

$$\frac{k_B T_0 r}{\hbar\sqrt{4\pi c}} = 1. \tag{33}$$

In this equation, T_0 implies one Prank emission.

When r in Eq. (33) is considered a wavelength, the source of this wavelength, λ, is the fluctuation energy gap, which is related to an e-neutrino self-energy [3]. The e-neutrino self-energy is expressed by the following equation [3]:

$$2\Delta_{e,\nu} = 0.025eV = 4.0 \times 10^{-21}J, \tag{124}$$

where $\Delta_{e,\nu}$ implies the energy level for an e-neutrino. Therefore, it is necessary to obtain a photon from this energy level. In this case, the product of α and this energy level creates photon energy gap $\hbar\omega$:

$$2(\Delta_{e,\nu} \times \alpha) = 2(\hbar\omega) = 4.0 \times 10^{-21}\alpha. \tag{125}$$

Thus, ω and λ are calculated as.

$$\omega = 1.9 \times 10^{13} \times \frac{1}{137} = 1.3 \times 10^{11}\ \text{rad/s} \tag{126}$$

and

$$\lambda = \frac{c}{\omega} = 2.3 \times 10^{-3}\text{m}. \tag{127}$$

The derived λ is substituted in Eq. (33) to obtain

$$T_0 = \frac{\hbar c\sqrt{4\pi}}{k_B \lambda} \approx 3.7\text{K}, \tag{128}$$

which agrees with the CMB temperature provided in [15, 16].

3.3 Depiction of a gravitation wave (chirp signal)

The derived equations from the theory section are again

$$h_{max} = \frac{t_c}{\Delta t} \tag{45}$$

and

$$u_p = \frac{t_c}{\Delta t} cos\,(\omega t_{00}), \tag{46}$$

where t_{00} is defined as a constant of 1[s] because ω is dependent on Δt.

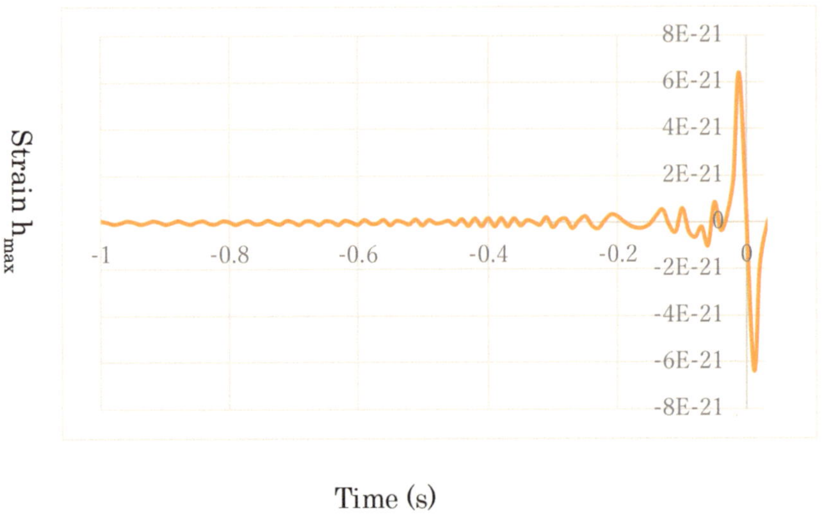

Figure 5.
Analytical calculation for the gravity wave. This figure was cited from [10]. The critical point indicates 10^{-21} order and the timescale is 0.1 s. The calculation agrees well with measurements and was obtained from the QEG equation solution.

Figure 5 shows the result of this analytical calculation of the gravity wave, which agrees with measurements provided in [11]. Considering that the strain $h_{\max}$ implies quantized space–time t_c, the gravity wave is a universal phenomenon. Moreover, $h_{\max}$ was derived with $z = 0$ of the QEG equation solution and, thus, the gravity wave has an anisotropic property [17].

3.4 Laws of electromagnetism derived from the QEG equation solution

3.4.1 Coulomb interaction and Poisson equations

Let us consider the Coulomb interaction and the satisfaction of the continuity equation regarding charge density and current density. The Poisson equations derived in the Theory section were

$$\frac{d^2\Phi}{dr^2} = \frac{-1}{\varepsilon_0}\frac{c^4}{16\pi G}\frac{1}{v}\frac{e}{m_e\omega_c}\frac{1}{n}r. \tag{60}$$

Thus, the charge density is

$$\rho = \frac{c^4}{16\pi G}\frac{1}{v}\frac{e}{m_e\omega_c}\frac{1}{n}r. \tag{60-2}$$

Eq. (60) can then be expressed as

$$\frac{d^2\Phi}{dr^2} = \frac{-1}{\varepsilon_0}\frac{c^4}{16\pi G}\frac{e}{m_e\omega_c^2}\frac{1}{n}. \tag{62}$$

For a vector potential,

$$\frac{d^2 A_\varphi}{dr^2} = \mu_0 \frac{c^4}{16\pi G} \frac{1}{n} \frac{e}{m_e \omega_c} r. \tag{69}$$

From Eq. (62), the Poisson equation for the electrostatic potential is given by

$$\rho_0 = \frac{c^4}{16\pi G} \frac{e}{m_e \omega_c^2} \equiv \beta_0 e, \tag{129}$$

where $n = 1$.
Generally,

$$\frac{dQ}{dv} = \beta_0 e, \tag{130}$$

where dv denotes the volume difference and

$$dQ \equiv e. \tag{131}$$

Thus,

$$\beta_0 = \frac{1}{dv} = \delta\left(\vec{r}\right). \tag{132}$$

Therefore, considering Eq. (62) and, as every elementary physics text states, the standard Coulomb potential forms as:

$$\Phi = \frac{e}{4\pi\varepsilon_o r}. \tag{133}$$

Note that r here is the relative distance between two charged particles because of the introduction of the Dirac function with position vector $\boldsymbol{r}$.

Next, we consider the satisfaction of the continuity equation by first considering the following elementary equation,

$$v \equiv \frac{dr}{dt}. \tag{134}$$

From Eq. (69), the current density is

$$i = \frac{-c^4}{16\pi G} \frac{1}{n} \frac{e}{m_e \omega_c} r. \tag{135}$$

Thus, the following equation is satisfied:

$$\frac{d\rho}{dt} + \frac{di}{dr} = 0. \tag{136}$$

Eq. (136) can be generalized to three dimensions as

$$\frac{\partial \rho}{\partial t} + div\,\vec{\imath} = 0. \tag{137}$$

Considering the satisfaction of both Eqs. (133) and (137), the Poisson equation for a vector potential is automatically proved. This is because the charge density and continuity equations have been proved. The current density, i.e., the Poisson equation for a vector potential as in Eq. (69), has also been proved.

3.4.2 Newtonian equation

Let us consider the energy of quantized space time in terms of the magnetic field (or gravity field).

In the Theory section, we derived

$$\Phi_G/2 = \frac{-1}{8\pi}\frac{1}{16\pi G}\frac{c^2}{m_e}g\frac{1}{r}. \tag{78}$$

Note that the 1/2 implies the symmetry of the flux direction of the magnetic field is broken, as considered in **Figure 1**. Using Eq. (78), we can calculate the gravity energy of a quantized space time in terms of magnetic or gravity field. If Φ_G has the unit [J], then parameter g has the unit $[J \bullet m^6]$. Thus, for **Figure 1** and considering the flux of the central circle,

$$g \equiv 1 \bullet \lambda_c^6 \approx 1 \bullet \lambda_0^6 \tag{138}$$

and in Eq. (78)

$$r \equiv \lambda_c \approx \lambda_0. \tag{139}$$

Because we are now calculating the energy of the magnetic field quantized space time in **Figure 1** (i.e., not the quantized space–times in terms of electric fields having embedded electrons, which have rest energy), the remaining energy among factor g is assumed to be the unit number one, which indirectly implies the existence of dark energy. Then, the potential is.

$$\Phi_G = -5.8 \times 10^{-10}\ \mathrm{J} \tag{140}$$

which is approximately equal to U_B in Eq. (117).

Using Eq. (118),

$$M_\tau = 1.8GeV/c^2. \tag{141}$$

Thus, Eq. (141) is approximately the remaining energy of a τ-particle [9] and also implies the energy of a quantized space time in terms of magnetic or gravity field. This value agrees with reported measurements and does not contradict the theory of this paper.

Note that the above Newtonian equation has a different shape from the standard Newtonian gravity equation usually taught in high school. However, although the standard Newtonian gravity equation is applied in the scale of the solar system, it is unnatural to consider that it can be applied on the quantum scale because every physics equation generally has application scales. For example, the equation $ma = F$ is well applied in macroscopic scales but cannot be applied in scales less than an atomic one. Thus, the success of Schrodinger's equation in application to the H atom comes not from the fact that the value of the standard Newtonian gravity equation is too small but from the fact that it is already considered inapplicable to the atomic scale.

3.4.3 Derivation of the time-dependent Maxwell's equations

Using the solution of the QEG equation,

$$r_i^2 \equiv (-2)\frac{8\pi G}{c^4}\frac{\frac{1}{2}\hbar\omega_i}{\lambda_c^3} + \left(c\frac{2\pi}{\omega}\right)^2, \tag{30}$$

we converted divisions of quantized space–times λ_c and t_c into standard differentials [1] to ensure the mathematical equation in the Theory section. Considering the Lorentz conservation regarding differentials, we derived the following equations:

$$rot\vec{H} = \frac{\partial\vec{D}}{\partial t} \tag{101}$$

and

$$\frac{\partial\vec{B}}{\partial t} = -rot\vec{E}. \tag{116}$$

Therefore, we claim that the above two equations are the same.

4. Discussion

4.1 Summary of the key points of this study

By introducing quantized space times derived from the zero-point energy, electromagnetic and gravity fields, including dark energy, are analytically well explained using the QEG equation. To this point, the only relevant concepts are zero-point energy and conservation of angular momentum of quantized space times.

4.2 Analytical solution to the QEG equation

The analytical solutions of the QEG equation resulted in various significant results. First, quantizing Einstein's gravity equation enables us to obtain the analytical (not numerical) solution that describes every electromagnetic and gravity field uniformly. According to our previous paper [1, 2], weak and strong interactions are essentially equal to static electromagnetic fields with consideration of the zero-point energy. Thus, this paper reinforces the results of our previous paper [1, 2], which describes unified field theory in terms of particle physics while indicating that the only source of every field is the zero-point energy. Moreover, the QEG equation solutions effectively describe existing phenomena in terms of the universe.

4.3 CMB emission

The analytical solution of the QEG equation also describes CMB emission. This result implies that we are not employing the standard big-bang model. We derived

CMB emission and a unique angular frequency that is a result of the QEG equation solution. The significance is that emission and absorption of CMB photons occur everywhere in our universe, and these emissions are directly related to the e-neutrino self-energy, which fluctuates in the energy level of the vacuum. Thus, CMB can be described without the standard big-bang model, and we thus claim that the measured CMB does not have the meaning of the initial time of birth of the universe.

4.4 Unified field in terms of electromagnetic and gravity fields

The analytical solution of the QEG equation also describes the unified field in terms of electromagnetic and gravity fields. This solution implies rotation of quantized space times both in terms of an electric field and a magnetic field (gravity field). The results lead to the Poisson equations regarding electrostatic, vector, and gravity potentials. These equations result in the Coulomb equation, Biot-Savart's law, which is derived from the Poisson equation for vector potential, and the Newtonian gravity equation.

In terms of the quantized space times, induction from both electric field to magnetic field and magnetic field to electric field are derived. Thus, the time-dependent Maxwell's equations are described. In short, the existing Einstein's gravity equation already contains properties of both electromagnetic and gravity fields. Thus, we claim that to obtain the unified field theory, it is not necessary to expand the existing Einstein's gravity equations, such as in five dimensions.

The most important point of this work is that all equations from electromagnetic and gravity fields come from the conservation law of angular momentum in terms of quantized space times. As mentioned in our previous paper [1, 2], weak and strong interactions are equal to electromagnetic fields and, thus, most microscopic fields and basic equations stem from the conservation law of angular momentum in terms of quantized space times. That is, only the zero-point energy is the source needed to create most fields.

Furthermore, the result of the analytical solution of the QEG equation automatically leads to the analytical derivation of gravity waves. The significance of this is that, although thus far gravity waves have only been obtained from numerical analysis of the existing Einstein's gravity equation, we have now derived them from the pure analytical solution of the QEG equation. This comes from the fact that we succeeded in the quantization of Einstein's gravity equation.

4.5 Three generations of leptons

Considering the basic configuration, including quantized space times in terms of both electric field and magnetic (gravity) field and the collapse of this configuration, we derived rest energies of both τ- and μ-particles that agree with measurement values. Considering that the real electron is the result of the collapse of the quantized spacetime configuration, we have now succeeded in providing the reason why leptons have three generations. The concept of quantized space times, in terms of electric, magnetic, or gravity field with zero-point energy, can be proven by comparison with measurements. In our previous paper regarding neutrinos [3], we described the three generations of neutrino, i.e., the oscillation of neutrinos and their self-energy, under the assumption that the masses of the three leptons are known in advance. However, we have now clarified all masses of the three leptons without assumption, and the most important mystery of why elementary particles have three generations was uncovered.

5. Conclusion

With the introduction of quantized space times derived from the zero-point energy and their conservation of angular momentum, i.e., the analytical solution of the QEG equation, we have created most laws and equations in terms of electromagnetic and gravity fields. Moreover, the configuration of quantized space times provides the reason why leptons have three generations.

The solution of the QEG equation also resulted in what is referred to as dark energy and the analytical derivation of gravity waves, which all agree well with reported measurements.

Conclusively, in this chapter, the gravitational wave was obtained using analytical calculations. Until now, this was only obtained using numerical or fitting methods.

With the combination of the results from our previous paper [1, 2], we have reinforced a unified field theory in terms of particle physics that indicates that concepts of zero-point energy and quantized space times describe most fields (i.e., electromagnetic field, gravity field, weak interaction, strong interaction, leptons, neutrinos, quarks, protons, neutrons, and so on). We selected the zero-point energy (i.e., the basic configuration of quantized spacetimes) as the basic source that describes almost all fields, including the masses of W and Z bosons. However, there is also the Higgs boson, which has not been described here or in our previous work. As a follow-up, it is necessary to achieve a consistent description that includes this boson.

Acknowledgements

1. We thank Enago (www.enago.jp) for their many times English language reviews.

2. We appreciate that the preprint version for this chapter can be accessed as Ref. [10].

Notation

The preprint version exists for this chapter. Please see Ref. [10]. There are some similarities as this preprint. However, these sources are from my original paper which is available online as preprint. However, in the preprint, there are a few careless errors in writing equations, and thus these have been modified in this chapter. Moreover, the logical flow in every portion of the preprint was checked and was revised in this chapter.

Additional information

This paper is not related to any competing interests such as funding, employment, and personal or financial interest. Moreover, this paper is not related to any non-financial competing interests.

Author details

Shinichi Ishiguri
Nihon University, Chiba, Japan

*Address all correspondence to: shinichi.ishiguri@gmail.com

References

[1] Ishiguri S. A unified theory of all the fields in elementary particle physics derived solely from the zero-point energy in quantized spacetime. Preprints. 2019:2019070326. DOI: 10.20944/preprints201907.0326.v1

[2] Ishiguri S. Studies on quark confinement in a proton on the basis of interaction potential. Preprints. 2019: 2019020021. DOI: 10.20944/preprints201902.0021.v1

[3] Ishiguri S. Theory on neutrino self-energy and neutrino oscillation with consideration of superconducting energy gap and Fermi's Golden rule. Preprints. 2019:2019100080

[4] Neronov A, Vovk I. Evidence for strong extragalactic magnetic fields from fermi observations of TeV blazars. Science. 2010;**328**:73

[5] Tavecchio F et al. The intergalactic magnetic field constrained by Fermi/ Large Area Telescope observations of the TeV blazar 1ES 0229+200. Monthly Notices of the Royal Astronomical Society. 2010;**406**:L70

[6] Riess AG et al. Observational evidence from supernovae for an accelerating universe and a cosmological constant. Astronomy Journal. 1998;**116**:1009

[7] Kawabata T et al. Time-reversal measurement of the p-wave cross sections of the 7Be(n,α)4He reaction for the cosmological Li problem. Physical Review Letters. 2017;**118**:052701

[8] Ishiguri S. Analytical descriptions of high-Tc cuprates by introducing rotating holes and a new model to handle many-body interactions. Preprints. 2020: 2020050105. DOI: 10.20944/ preprints202005.0105.v1

[9] Hara Y. Elementary Particle Physics. Shokabo in Tokyo; 2003. p. 191

[10] Ishiguri S. Unified field theory for electromagnetic and gravity fields with the introduction of quantized space–time and zero-point energy. Preprints. 2020:2020070462. DOI: 10.20944/ preprints202007.0462.v1

[11] Abbott BP. GW150914: The advanced LIGO detectors in the era of first discoveries. Physical Review Letters. 2016;**116**:061102

[12] Cutler C et al. The last three minutes: Issues in gravitational-wave measurements of coalescing compact binaries. Physical Review Letters. 1993;**70**:2984

[13] Ishiguri S. New attractive-force concept for Cooper pairs and theoretical evaluation of critical temperature and critical-current density in high-temperature superconductors. Results in Physics. 2013;**3**:74

[14] Ishiguri S. Potential of new superconductivity produced by electrostatic field and diffusion current in semiconductor. Journal of Superconductivity and Novel Magnetism. 2011;**24**(1):455-462

[15] Bennett CL et al. Nine-year wilkinson microwave anisotropy probe (wmap) observations: Final maps and results. The Astrophysical Journal. 2013;**208**

[16] Dicke RH, Peebles PJE, Roll PG, Wilkinson DT. Cosmic black-body radiation. The Astrophysical Journal. 1965;**142**:414

[17] Maggiore M. Gravitational Waves, Volume 1: Theory and Experiments. Gravitational Waves. Oxford: Oxford University Press; 2007

Chapter 4

New Conservation Law as to Hubble Parameter, Squared Divided by Time Derivative of Inflaton in Early and Late Universe, Compared with Discussion of HUP in Pre Planckian to Planckian Physics, and Relevance of Fifth Force Analysis to Gravitons and GW

Andrew Walcott Beckwith

Abstract

We use as given by Li and Koyama, as to using their idea of a fifth force. In doing so, we are assuming a force which is minus the spatial derivative of a scalar field. The scalar field we are using is one from Padmanabhan, and the problem is that the scalar field in the Padmanabhan representation is initially only dependent on time. The time component is stated to be in the Pre Planckian regime is proportional to a radial distance divided by the speed of light. The rephrasing of time as justified by stating that time in its initial configuration does not exist before the expansion of the universe and that we reintroduce time separately from a radial component divided by the speed of light upon entrance into Planckian space–time. In doing this we also refer to a new assumed conservation law which will give new structure as to inflationary expansion and its immediate aftermath. That of the Hubble "constant" divided by the 'time derivative' of the scalar field in the inflation regime and then a long time afterwards.

Keywords: inflaton, fifth force, gravitational waves, gravitons, Hubble parameter

1. Introduction

Our idea is to regularize inflation and its aftermath by a Hubble parameter divided by the derivative of a scalar field, as being about the same ratio in Planckian space time and then say in the time frame within a billion years of the present. The benefits of such an interpretation is to regularize how we obtain GW frequency, in the initial phase of the universe expansion to near the present era. In addition we use a fifth force as to explain how we would have almost an infinite expansion rate in the beginning of inflation. The almost infinite expansion rate is due to the fifth force which triggers rapid expansion. We then conclude as to a regime of black hole physics, with a table as to a pre present universe well before our big bang super massive black hole, which would then be through a nonsingular start to the universe break up from pre big bang configuration into millions of micro sized black holes. The way to do this assumes a variant of the Penrose cyclic conformal cosmology model, with a pre universe giant sized black holes broken up into tiny black holes by the uncounted millions [1, 2].

2. How we will obtain scalar field behavior we want which yields input into a fifth force

Using

$$a(t) = a_{initial} t^{\nu} \tag{1}$$

Which will lead to

$$\frac{H^2}{\dot{\phi}} \approx \sqrt{\frac{4\pi G}{\nu}} \cdot t \cdot T^4 \cdot \frac{(1.66)^2 \cdot g_*}{m_P^2} \approx 10^{-5} \tag{2}$$

The Eq. (2) is a conservation law which is considered to be true in the initial expansion, Planck regime of space–time.

This of course makes uses of Eq. (3) for the Hubble parameter, the Padmabhan value of the scalar field due to Eq. (1) and this is all assuming a value of

$$H = 1.66\sqrt{g_*} \cdot \frac{T_{temperature}{}^2}{m_P} \tag{3}$$

We will make the following calculation [3, 4].

$$V_0 = \left(\frac{.022}{\sqrt{qN_{efolds}}} \right)^4 = \frac{\nu(\nu - 1)\lambda^2}{8\pi G m_P^2} \tag{4}$$

λ as a dimensionless parameter which we refer to later. From [3], page 17 we have ae\ Chamelon mechanism for fifth force as

$$F_{5th-force} = -\frac{\tilde{\beta} \cdot \left(\vec{\nabla} \phi \right)}{m_P} \tag{5}$$

Eq. (5) equals zero of we have a scalar field solely dependent upon time. i.e. we need to have a re set of time as initially spatial divided by the speed of light.

3. Pre-Planckian to Planckian regime of space-time reset so Eq. (5) for fifth force does not vanish

To do this we will assume in an initial "bubble' of space–time of which we have spatial r value and a speed of light given by c

$$t = \frac{r}{\varpi c} \tag{6}$$

The term of ϖ is a dimensionless value and never negative.

If so, Eq. (5) will yield [3, 4].

$$F_{5th-force} = -\frac{\tilde{\beta} \cdot \left(\vec{\nabla}\phi\right)}{m_P} \approx -\frac{\tilde{\beta}}{2m_P r} \cdot \sqrt{\frac{\nu}{\pi G}} \tag{7}$$

4. What is the power of production of gravitons due to fifth force?

The easiest way is to look at power expressions for GW and to make them linked to Eq. (7).

First Power = Force, time velocity.

$$P = Power = F(force) \times v(velocity) \tag{8}$$

Compare Eq. (8) for Power in terms of gravitational waves using [5–7].

See [7],

$$P_{GW} \approx \frac{GM_{mass}\omega_{gw}^2}{c^2} \tag{9}$$

Usually, we want to look at GW quadrupoles, [6], page 312

$$\ddot{Q}^2 \approx \left[\left(\frac{\omega_{gw}^2}{c^2}\right) \cdot \langle r^2 \rangle\right]^2 \tag{10}$$

Keep in mind that we are using GW power which is given by

$$P_{GW} \approx \frac{Gc \cdot (M_{mass})^2 \omega_{gw}^6 \langle r^2 \rangle^2}{c^6} \tag{11}$$

Eq. (7) times c (speed of light) will alter Eq. (11) to be read as

$$\begin{aligned} P_{GW} &\approx \frac{Gc \cdot (M_{mass})^2 \omega_{gw}^6 \langle r^2 \rangle^2}{c^6} \\ &\approx c \times |F_{5th-force}| = \left|-c \times \frac{\tilde{\beta} \cdot \left(\vec{\nabla}\phi\right)}{m_P}\right| \approx c \times \frac{\tilde{\beta}}{2m_P r} \cdot \sqrt{\frac{\nu}{\pi G}} \end{aligned} \tag{12}$$

This leads to

$$\omega_{gw}^{6} \approx c^{7} \times \frac{\tilde{\beta}}{2m_{P}r} \cdot \sqrt{\frac{\nu}{\pi G}} \times \frac{1}{Gc \cdot (M_{mass})^{2} \langle r^{2} \rangle^{2}}$$
$$\Rightarrow \omega_{gw} \approx \left(\sqrt{\frac{\nu}{4\pi G}} \times \frac{\tilde{\beta} \cdot c^{6}}{G \cdot (M_{mass})^{2} m_{P}r \cdot \langle r^{2} \rangle^{2}} \right)^{1/6} \tag{13}$$

In terms of Planck Units

$$\omega_{gw}^{6} \approx c^{7} \times \frac{\tilde{\beta}}{2m_{P}r} \cdot \sqrt{\frac{\nu}{\pi G}} \times \frac{1}{Gc \cdot (M_{mass})^{2} \langle r^{2} \rangle^{2}}$$
$$\Rightarrow \omega_{gw} \approx G, m_{P}, r \approx \ell_{P} \xrightarrow[Planck-normalization]{} 1$$
$$M_{mass} \approx \varsigma \cdot m_{P} \xrightarrow[Planck-normalization]{} \varsigma \tag{14}$$
$$\langle r^{2} \rangle^{2} \approx \ell_{P}^{4} \xrightarrow[Planck-normalization]{} 1$$
$$\therefore \omega_{gw} \xrightarrow[Planck-normalization]{} \left(\sqrt{\frac{\nu}{4\pi}} \times \frac{\tilde{\beta}}{(\varsigma)^{2}} \right)^{1/6}$$

We find then we have at the immediate beginning of inflation, an almost Planck frequency value of 1.855 times 10^43 Hertz, we would need ν be 10^502 which would be factored into Eq. (1) and the scale factor value for the term ν. This would mean for the fifth force argument that we would have an almost infinitely quick expansion in the neighborhood of Planck length for the start of inflation.

What this means is that coefficient ν in the initial genesis of GW which will be in Planckian space–time to be

$$\nu \xrightarrow[Planck-normalization]{} 4\pi \times \left(\omega_{gw}\right)^{12} \times \frac{(\varsigma)^{4}}{\tilde{\beta}^{2}} \tag{15}$$

If we are looking at Planck time, in the Planck era, $\nu \propto \left(\omega_{Planck}\right)^{12}$, meaning that the rate of expansion in the early universe is commensurate with inflation.

5. Interpreting force assumed in terms of Ehrenfests theorem

Gasiorowitz, [5] gives this Ehrenfests Theorem as

$$F = \frac{d\langle p \rangle}{dt} = -\left\langle \frac{dV}{dr} \right\rangle_{t} \tag{16}$$

We re write Eq. (16) as yielding in our procedure

$$\langle p \rangle = -\frac{\tilde{\beta}}{2m_{P}} \cdot \sqrt{\frac{\nu}{\pi G}} \cdot \frac{\ln t}{\varpi c} \tag{17}$$

For sufficiently small time step, t, use the ideas given in, [5] which leads to

$$\begin{aligned} & \text{If } \langle p\rangle \approx \Delta p \approx \frac{\tilde{\beta}}{2m_P}\cdot\sqrt{\frac{\nu}{\pi G}}\cdot\frac{|\ln\varepsilon^{+}|}{\varpi c} \\ & \Delta p\Delta x \approx \hbar \Rightarrow \Delta x \approx \frac{\hbar}{\frac{\tilde{\beta}}{2m_P}\cdot\sqrt{\frac{\nu}{\pi G}}\cdot\frac{|\ln\varepsilon^{+}|}{\varpi c}} \leq l_P \end{aligned} \tag{18}$$

Meaning we will be obtaining to enormous energy values, for time smaller than Planck time.

6. Energy values, and the degrees of freedom initially

In an earlier paper, initial mass [8] is written as a huge value, namely

$$\begin{aligned} M &= \sqrt{\sqrt{g_*}\cdot\frac{1.66\hbar}{64\pi^2 m_P G^2 k_B^2}}\cdot\sqrt{\frac{t}{\gamma}}\sqrt{N_{Gravitons}}\cdot m_{Planck} \\ & \xrightarrow[Planck-Units]{} \approx \sqrt[4]{g_*}\cdot\sqrt{\frac{1.66}{64\pi^2}}\cdot m_{Planck} \approx \sqrt{N_{Gravitons}}\cdot m_{Planck} \\ & \approx 10^{60}\cdot m_{Planck} \end{aligned} \tag{19}$$

If so then

$$\sqrt[4]{g_*}\cdot\sqrt{\frac{1.66}{64\pi^2}} \approx 10^{60} \tag{20}$$

i.e. the initial degrees of freedom, would be

$$g_* \approx 10^{240}\cdot\left(\frac{64\pi^2}{1.66}\right)^2 \approx 10^{240}\times 144791 \propto 10^{245} \tag{21}$$

The magnitude of Eq. (21) is a ten to the 245 value, whereas degrees of freedom, is 10 to the 61st power.

Why we pay attention to this value. By [8].

$$\begin{aligned} & m_{graviton} \approx 10^{-60} m_P \Rightarrow N_{Gravitons} \approx 10^{120} \\ & \Rightarrow N_{Gravitons} \approx 10^{120} \approx S_{entropy} \\ & \Leftrightarrow g_* \approx 10^{240}\cdot\left(\frac{64\pi^2}{1.66}\right)^2 \approx 10^{240}\times 144791 \propto 10^{245} \end{aligned} \tag{22}$$

This is directly due, if assuming, an initial value of $t = \frac{r}{\varpi c}$ that Eq. (5) would be a large negative value, and

$$\phi\left(\frac{r}{\varpi c}\right) = \sqrt{\frac{\nu}{4\pi G}}\ln\left(\sqrt{\frac{8\pi G V_0}{\nu(\nu-1)}}\cdot\left(\frac{r}{\varpi c}\right)\right) \tag{23}$$

We will next discuss the fifth force in terms of the Dilaton model.

7. Dilaton model versus the general relativity theory. Can we extend general relativity?

In [3], page 17, the NON relativistic geodestic equation for a 'test particle'

$$\ddot{\vec{x}} = -\vec{\nabla}\Psi - \frac{\tilde{\beta}\cdot\left(\vec{\nabla}\phi\right)}{m_P} \tag{24}$$

The first term is gravitational potential Ψ. The second term is the fifth force. For Pre Planck to Plank

$$\ddot{\vec{x}} = -\vec{\nabla}\Psi - \frac{\tilde{\beta}\cdot\left(\vec{\nabla}\phi\right)}{m_P} \xrightarrow[\text{Pre}-Planck]{} \ddot{\vec{x}} = -\frac{\tilde{\beta}\cdot\left(\vec{\nabla}\phi\right)}{m_P} \tag{25}$$

We assume

$$\begin{aligned}
&-\vec{\nabla}\Psi \xrightarrow[\text{Pre}-Planck]{} 0 \\
&-\frac{\tilde{\beta}\cdot\left(\vec{\nabla}\phi\right)}{m_P} \xrightarrow[\text{Pre}-Planck]{} NOTzero \\
&t \xrightarrow[\text{Pre}-Planck]{} t = \frac{r}{\varpi c} \\
&t \xrightarrow[Planck]{} t \neq \frac{r}{\varpi c} \\
&-\frac{\tilde{\beta}\cdot\left(\vec{\nabla}\phi\right)}{m_P} \xrightarrow[Planck]{} Very - small - value \\
&-\vec{\nabla}\Psi \xrightarrow[Planck]{} Not - zero
\end{aligned} \tag{26}$$

In addition [1–4] we assume where we are using the values of an inflation potential given in [2] if we have Eq. (1) for the scale factor compared with another similar scalar potential

$$V(\phi) = V_0 \exp\left(-\frac{\lambda\phi}{m_P}\right) \leftrightarrow V_0 \exp\left(-\sqrt{\frac{16\pi G}{\nu}}\cdot\phi\right) \tag{27}$$

In Pre Planck physics, Eq. (5) would be enormous, whereas the fifth force for Plank regime and beyond would be small. Also the Gravitational physics term due to a gravitational potential Ψ would be the largest term in Eq. (24).

8. Review of what our presumption of pre Planck to plank physics have gained us, before Gravimagnetism

We deviate from standard relativity and Newtonian physics by the existence of a fifth force in Pre Planckian to Planckian space–time physics.

We are considering what if Eq. (5) and Eq. (6) insert fifth force physics into cosmology What has to be determined are experimental verifications of Eq, (23) and Eq. (24). This is a test and a way to obtain falsifiable models. Furthermore we are presuming a nonsingular start to the universe. And these ideas need experimental verification.

Finally the model included in by use of references [9, 10] need to be seriously reviewed.

9. Gravimagnetism and an invariance model considered

We revisit Gravimagnetism and its links to this problem [11].

Note on page 48 of [11].

$$\begin{aligned} &\frac{d\vec{v}}{dt} \equiv -grad\phi + 2\vec{\Omega} \times \vec{v} \\ &\leftrightarrow Lorentz - force \\ &= \vec{K} = q \cdot \left(\vec{E} + \frac{\vec{v}}{c} \times \vec{B} \right) \end{aligned} \tag{28}$$

In Electromagnetic theory we note we have exact correspondences. However in GR the first line of Eq. (28) is approximate.

In the Pre Planck regime of space–time we use the following

$$\frac{d\vec{v}}{dt} \equiv -grad\phi + 2\vec{\Omega} \times \vec{v} \tag{29}$$

In the Pre Planckian regime we will have

$$\frac{d\vec{v}}{dt} \equiv -grad\phi + 2\vec{\Omega} \times \vec{v} \xrightarrow[\text{Pre}-Planckian]{} \frac{d\vec{v}}{dt} \equiv -grad\phi \approx -\partial_r \phi \tag{30}$$

Whereas we use this substitution to obtain a nonzero fifth force in Pre Planckian physics

$$\phi(t) \xrightarrow[\text{Pre}-Planckian]{} \phi\left(\frac{r}{\varpi c}\right) \tag{31}$$

If this is done, then the Graviton condensate relationship as argued by the author before, should also be examined as far as experimental verification, It would be optimal if we find that the Pre Planckian to Planckian physics regime would have a lot of black holes, as given in [12].

$$\begin{aligned} &m \approx \frac{M_P}{\sqrt{N_{gravitons}}} \\ &M_{BH} \approx \sqrt{N_{gravitons}} \cdot M_P \\ &R_{BH} \approx \sqrt{N_{gravitons}} \cdot l_P \\ &S_{BH} \approx k_B \cdot N_{gravitons} \\ &T_{BH} \approx \frac{T_P}{\sqrt{N_{gravitons}}} \end{aligned} \tag{32}$$

Having a change in initial conditions from Pre Planckian physics to Planckian physics would be enough if we find that m in Eq. (32) is actually the mass of a graviton.

If so, by Novello [13] we then scale mass m as given in Eq. (32) to the mass of a graviton, as in Eq. (33)

$$m_g = \frac{\hbar \cdot \sqrt{\Lambda}}{c} \tag{33}$$

In a word, the next step to ascertain would be how Eq. (31), as given breaks down, and we have then application of Eq. (32) with m set with m becoming the mass of a graviton as given in Eq. (33).

Confirming these details should be the object of future research as can also be seen in [13].

In addition we have the [14] as to the choice of the Starobinsky potential as well use of radial acceleration as a way of confirming the cosmological constant.

The way indicated in [14] may fix the value of m, after determining M, as an input into Eq. (32).

Then use the right hand side of Eq.(33) whereas in [8], we will be determining the right hand side of Eq. (33), namely Λ and then after doing that, assuming Eq. (33) to work backwards into the M of Eq. (32).

That is how to reconcile the [8] and [14] references whereas we will be using this current document to ascertain the existences of a Fifth force which would be a bridge between Pre Planckian to Planckian physics,. Finally though what is implicitly assumed is [15] which is an application of Klauder enhanced quantization.

Finally is the imponderable, i.e. the generalization of Penrose CCC theory which is in [8] which is a generalization of what is in Penrose single universe recycling of universes which may be seen in [16].

All these steps need to be combined and rationalized. Also remember [17].

10. Now for the invariance model

Also the Hawking argument as to the probability of finding a universe with Λ being a given value [18, 19].

$$P(\Lambda) \sim \exp\left(-2S_E(\Lambda)\right) \approx \exp\left(\frac{3\pi M_P^2}{\Lambda}\right) \tag{34}$$

We get combining Eq. (32), Eq. (33) and Eq. (34) to realize having

$$\begin{aligned} P(\Lambda) &\xrightarrow[Eq(1),Eq.(2),Eq.(3)]{} \exp\left(\frac{3\pi c^2 N_{graviton}}{\hbar^2}\right) \\ &\xrightarrow[Eq(1),Eq.(2),Eq.(3),\hbar=c\to 1]{} \exp\left(3\pi N_{graviton}\right) \end{aligned} \tag{35}$$

11. Now putting in the detail about the universe being treated as a giant black hole, of sorts

First sign in the mass m in Eq. (32) as being the same as the mass of a graviton, in Eq. (33).

We then would have

$$m \to m_g \approx \frac{M_P}{\sqrt{N_{graviton}}} \Rightarrow N_{graviton} \approx 10^{122} \tag{36}$$

In addition the radius of the "particle" would be of the form given by

$$R \to R_{universe} \approx \sqrt{N_{graviton}} \cdot \ell_P \approx 10^{61} \cdot \ell_P \tag{37}$$

Also the overall mass M would scale as

$$M \to M_{universe} \approx \sqrt{N_{graviton}} \cdot M_P \approx 10^{61} \cdot M_P \tag{38}$$

Whereas the entropy

$$S \to S_{universe}(gravitons) \approx k_B \cdot 10^{122} \xrightarrow[k_B \to 1]{} 10^{122} \tag{39}$$

And the final temperature

$$T \to T_{universe}(gravitons) \approx \frac{T_P}{\sqrt{N_{graviton}}} \approx 10^{-61} \cdot T_P \tag{40}$$

In this case, we have that the mass of the graviton, allowing for this scaling is given by [5, 6, 20, 21].

12. Consequences. We have a starting point determined by the following

From Eq. (1) and Eq. (2) of this manuscript we have the DNA for the working out of the coefficient of the scale factor, and this is in the end what we end up with.

If we are looking at Planck time, and assuming we have Plank frequency, this means in the Planck era

$$\nu \propto (\omega_{Planck})^{12}, \tag{41}$$

This enormous initial coefficient to the scale factor time coefficient, would be put in initially in the last part of Eq. (41) which would subsequently, be invariant, namely from the beginning of inflation, to its near present day conditions, the following would be invariant, so the following would be approximately a constant

$$\left.\frac{H^2}{\dot{\phi}} \approx 10^{-5}\right\}_{initial-conditions} \xrightarrow[Evolution-to-near-present]{} \left.\frac{H^2}{\dot{\phi}} \approx 10^{-5}\right\}_{present-conditions} \tag{42}$$

This would somehow have to be confirmed via data sets but the coefficients in the initial conditions to final, in ratio would be similar, in ratio value, but the magnitude of the H term, and the magnitude of the derivative of the scalar field would be vastly different, just their ratios would likely have a similar value [22–25]. And this leads us to the final question to raise. What would it take to come up with the initial frequency as given in Eq. (41) leading to an initially extremely high rate of expansion of the scale factor? To see this let us conclude about energy initially,

13. So do we have something (space time) from nothing? Conclusion with speculations as to answer

To answer this, we look at the following. Namely the crazy geometry in the Pre Planckian regime of space time.

Let us first recall the Shalyt-Margolin and Tregubovich (2004, p.73) [26].

$$\Delta t \geq \frac{\hbar}{\Delta E} + \gamma t_P^2 \frac{\Delta E}{\hbar} \Rightarrow (\Delta E)^2 - \frac{\hbar \Delta t}{\gamma t_P^2} (\Delta E)^1 + \frac{\hbar^2}{\gamma t_P^2} = 0$$
$$\Rightarrow \Delta E = \frac{\hbar \Delta t}{2\gamma t_P^2} \cdot \left(1 + \sqrt{1 - \frac{4\hbar^2}{\gamma t_P^2 \cdot \left(\frac{\hbar \Delta t}{2\gamma t_P^2}\right)^2}} \right) = \frac{\hbar \Delta t}{2\gamma t_P^2} \cdot \left(1 \pm \sqrt{1 - \frac{16\hbar^2 \gamma t_P^2}{(\hbar \Delta t)^2}} \right) \tag{43}$$

For sufficiently small γ.

$$\Delta E \approx \frac{\hbar \Delta t}{2\gamma t_P^2} \cdot \left(1 \pm \left(1 - \frac{8\hbar^2 \gamma t_P^2}{(\hbar \Delta t)^2} \right) \right)$$
$$\Rightarrow \Delta E \approx either \frac{\hbar \Delta t}{2\gamma t_P^2} \cdot \frac{8\hbar^2 \gamma t_P^2}{(\hbar \Delta t)^2}, or \frac{\hbar \Delta t}{2\gamma t_P^2} \cdot \left(2 - \frac{8\hbar^2 \gamma t_P^2}{(\hbar \Delta t)^2} \right) \tag{44}$$

would lead to a minimal relationship between change in E and change in time as

$$\Delta E \approx \frac{\hbar \Delta t}{2\gamma t_P^2} \cdot \frac{8\hbar^2 \gamma t_P^2}{(\hbar \Delta t)^2} \equiv \frac{4\hbar}{\Delta t} \tag{45}$$

Or

$$\Delta E \Delta t \approx 4\hbar \tag{46}$$

In doing so, we will refer to Eq. (46) as the pre inflaton state of energy being delivered due to a non conserved interjection of energy into the new universe.

Doing so would be a way to have the frequency so alluded to given in [27–29] and this is what we conjecture as to the evolution of the change in energy if we have the inflaton included which would be in Planckian space–time

$$\left\langle (\delta g_{uv})^2 (\hat{T}_{uv})^2 \right\rangle \geq \frac{\hbar^2}{V_{Volume}^2} \xrightarrow[uv \to tt]{} \left\langle (\delta g_{tt})^2 (\hat{T}_{tt})^2 \right\rangle \geq \frac{\hbar^2}{V_{Volume}^2} \& \delta g_{rr} \sim \delta g_{\theta\theta} \sim \delta g_{\phi\phi} \sim 0^+ \tag{47}$$

$$\delta t \Delta E \geq \frac{\hbar}{\delta g_{tt}} \neq \frac{\hbar}{2}$$
$$Unless \delta g_{tt} \sim O(1) \tag{48}$$

$$\delta g_{tt} \sim a^2(t) - \phi < < 1 \tag{49}$$

This version of uncertainty would be for inclusion of energy once we are in the specific Planckian regime of space time and may be what is needed for sufficient energy imput from the fifth dimension, leading to a fifth force argument as given by [30] which may be from the work given by Wesson in [27]. This fifth force, in addition to fitting in the HUP in the Pre Planckian to Planckian physics regime would be encouraging us for an unbelievably high initial change in energy, as stated in Eq. (48), whereas once we are in the Planckian regime of the present universe we would be using Eq. (46) so as to specify a very high initial initial frequency, and this would be in tandeum with [27] being directly employed

$$\begin{aligned} &\int p_\alpha dx^\alpha = \pm \frac{h}{c} \cdot \frac{L}{\ell} = \pm \frac{h}{c} \cdot \sqrt{\frac{3}{\Lambda}} \frac{m_{particle}}{h} \\ &= \frac{r}{m_{pl}} \cdot \left(1 - \log\left[\frac{r}{\varpi \cdot c}\right]\right) \\ &\Leftrightarrow r \approx \varepsilon^+ \end{aligned} \tag{50}$$

This is linkable to z, as to red shift showing up in [27–29] and it shows how to obtain a very small radial value namely in a tiny scale factor due to an enormous z red shift as given in [29].

Quote.

$If z \approx 10^{55} then a \approx 10^{-55}$ so we do not have a space – time singularity.

End of quote.

The Eq. (46) would be for specifying, via the frequency being the inverse of change in time, after the Planckian regime of space time, whereas Eq. (47) would be the Pre Planckian to Planckian uncertainty principle used when Eq. (50) would be considered.

The application of the fifth force to the geometry of space–time in the beginning of expansion of the universe would employ Eq. (50) and Eq. (48) in the Planckian regime, whereas Eq. (47) would be just before Planckian space time.

All these details need to be worked out and given more foundation in future research.

14. Linkage to GW and their importance as to GW astronomy by making an analogy to black holes explicit for GW generation, and this has to be confirmed experimentally

We will for the sake of linkages to GW treat this problem as related to black holes, and gravitons and subsequent GW generation.

The Eq. (42) so mentioned, is an invariance procedure as far as space–time and its scaling may lead to black hole production'.

Assuming our BEC condensate argument leads to scaling as far as black hole production, we will make the following assumption, namely the following grouping leads to

$$\begin{aligned} &M \rightarrow M_{universe} \approx \sqrt{N_{graviton}} \cdot M_P \approx 10^{61} \cdot M_P \\ &M \rightarrow several \tilde{m} = \frac{8\pi R(radius - of - \tilde{m})^3 \tilde{\rho}}{3} \\ &\left.\frac{H^2}{\dot{\phi}} \approx 10^{-5}\right\}_{initial-conditions} \xrightarrow{Evolution-to-near-present} \left.\frac{H^2}{\dot{\phi}} \approx 10^{-5}\right\}_{present-conditions} \end{aligned} \tag{51}$$

I estimate that this together leads to about 10^20 to 10^21 effective Planck mass s sized mini black holes in the beginning of the cosmos at the cosmos. Making use of page 46 of [31] we have that 1/1000 of a 3 + 1 dimensional mini black hole would, if not considered rotating contribute to graviton emission.

Using that rule, we could assume 10^`122 gravitons, as actually being generated from primordial conditions with say of this number, say at most about 10^21 Planck sized black holes being formed.

Then

$$
\begin{aligned}
& E_{BEC-Graviton} \approx \frac{k_B T_{BH}}{2} \approx \frac{k_B \times 10^{-5} \times T_P}{2} \\
& \Rightarrow \omega_{BEC-Graviton} \propto 10^{-5} \times 10^{43} Hz \approx 10^{38} Hz \\
& \Rightarrow \omega_{BEC-Graviton-to-CMBR} \approx 10^{38} \times 10^{-3} Hz
\end{aligned} \tag{52}
$$

We could see Primordial black holes as about $z \approx 10^{25}$. Leading to present Gravitational wave signals from the primordial black holes today of about 1 Hertz, by massive red shifting.

Whereas we can consider what would be gained by looking at the contribution near the CMBR, i.e. z ~ 1100 or so for the CMBR,

whereas this would mean roughly that we would be looking in the regime of the CMBR generated processes

$$
\omega_{signal-from-Planck-to-CMBR} \propto \left(\frac{3}{2}\right)^{1/\gamma} \cdot 10^{(25/\gamma)} \times 10^{-3} Hz \tag{53}
$$

Also

$$
\Delta E \Delta t \approx \hbar \equiv \hbar \omega \Delta t \approx \hbar \omega \cdot \left(\frac{2}{3a_{\min}}\right)^{1/\gamma} \Rightarrow \omega \approx \hbar^{-1} \cdot \left(\frac{2}{3a_{\min}}\right)^{-1/\gamma} \tag{54}
$$

We claim that if we take the energy as consistent with a change in value as given by Eq. (45) and Eq. (48) that this will lead to a frequency which may, if $a_{\min} \approx 10^{-25} - 10^{-20}$ (range from 10^-25 to 10^-20) lead to initial primordial production of Frequencies as to emergence from a near singularity along the lines of an initial value of

$$
\omega \approx \hbar^{-1} \cdot \left(\frac{3}{2}\right)^{1/\gamma} \cdot 10^{25/\gamma} \propto \left(\frac{3}{2}\right)^{1/\gamma} \cdot 10^{25/\gamma} Hz \tag{55}
$$

Whereas as from [32] we assume the following table as given in that publication for say a huge number of initial primordial black holes.

This would lead to about an energy release initially of the order of say

$$
\begin{aligned}
& \dot{E} = GW - change - in - energy = \frac{32 (M_1 M_2)^2 (M_1 + M_2)}{5 \cdot R^2 M_{Planck}^5} \\
& \xrightarrow[M_1 = M_2 = M_{Planck}]{} \frac{64}{5 \cdot R^2 (Planck - length)} \\
& \equiv Change - in - power - from - Rotating - binary - blackholes
\end{aligned} \tag{56}
$$

End of Prior Universe time frame	**Mass (black hole): super massive end of time BH 1.98910^+41 to about 10^44 grams**	**Number (black holes) 10^6 to 10^9 of them usually from center of galaxies**
Planck era Black hole formation Assuming start of merging of micro black hole pairs	Mass (black hole) 10^-5 to 10^-4 grams	Number (black holes) 10^40 to about 10^45,
Post Planck era black holes	Mass (black hole) Up to 10^6 grams per black hole	Number (black holes) 10^20 to at most 10^25

Table 1.
Black hole production per cosmological era.

We of course would be wanting to compare this with the change in energy as given in Eq. (45) and Eq. (48).

Table 1 from [32] assuming Penrose recycling of the Universe as stated in that document.

Now for the sake of primordial black holes.

The formula in page 16 of reference [32] that two black holes emit GW with a wave frequency 2 times the rotation frequency of the orbit of the two black holes to each other.

If we assume that we are still using this approximation above, from [33].

$$\begin{aligned} R(separation) \simeq r_g^{eff} &= \frac{(M_1 + M_2)}{(M_{Planck})^2} \\ \xrightarrow[M_1 = M_2 = M_{Planck}]{} \xrightarrow[M_{Planck}=1]{} & 1 \equiv R(Planck - length) \end{aligned} \tag{57}$$

I.e. this means that the primordial black holes, presumably of Planck size would be separated about 1 Planck length from each other, that their recombination would be quick and that the frequency range would likely be of the magnitude of about 10^25 Hz in terms of GRAVITATIONAL waves which would then be massively red shifted downward to about `1 Hz in an Earth bound detector system.

I.e. a huge downward red shifting from 10^25 Hz to about a 1 Hz value in Earth orbit.

Acknowledgements

This work is supported in part by National Nature Science Foundation of China grant No. 11375279.

Author details

Andrew Walcott Beckwith
Physics Department, Chongqing University, People's Republic of China

*Address all correspondence to: rwill9955b@gmail.com

References

[1] Sarkar U. Particle and Astroparticle Physics. New York City, New York, USA: Taylor and Francis; 2008

[2] Padmanabhan T. An invitation to astrophysics. In: World Scientific Series in Astronomy and Astrophysics. Vol. 8. Singapore, Republic of Singapore: World Press Scientific; 2006

[3] Li B, Koyama K. Modified Gravity. Hakensack New Jersey, USA: World Scientific; 2020

[4] Dimopoulos K. Introduction to Cosmic Inflation and Dark Energy. Boca Raton, Florida, USA: CRC press; 2021

[5] Gasiorowitz S. Quantum Physics. New York City, New York, USA: John Wiley and Sons; 1974

[6] Walecka J. Introduction to General Relativity. Hackensack, New Jersey, USA: World Scientific; 2007

[7] Lightman A, Press W, Price R, Teukolsky S. Problem Book in Relativity and Gravitation. Princeton, New Jersey, USA: Princeton University Press; 1975

[8] Beckwith A. How Initial Degrees of Freedom May Contribute to Initial Effective Mass, i.e. Effective Mass of the Universe Proportional to (D.O.F.) to the 1/4th Power by an Enormous Initial Degree of Freedom Value. Journal of High Energy Physics, Gravitation and Cosmology. 2022;**8**:1127-1133. DOI: 10.4236/jhepgc.2022.84079

[9] Ng Y. Spacetime Foam: From Entropy and Holography to Infinite Statistics and Nonlocality. Entropy. 2008;**10**(4): 441-461. DOI: 10.3390/e10040441

[10] Ruutu V, Eltsov V, Gill A, Kibble T, Krusius M, Makhlin YG, et al. Vortex formation in neutron – Irradiated ^{3}He as an analog of cosmological defect formation. Nature. 1996;**382**:334-336

[11] Jetzer P. Applications of General Relativity, with Problems. Cham, Swsitzerland: Springer Verlag; 2022

[12] Chavanis, P. "Self Gravitating Bose-Einstein Condensates", pp. 151-194, Quantum Aspects of Black Holes, Calmet, X, Editor, Fundamental Theories of Physics, 178, Springer Nature, Cham, Switzerland, 2012

[13] Novello M. The mass of the graviton and the cosmological constant puzzle. In: https://arxiv.org/abs/astro-ph/0504505, and In: Fifth International Conference on Mathematical Methods in Physics – IC 2006 April 24–28 M Centro Brasileiro de Pesquisas Fisicas. Rio de Janeiro, Brazil; 2006, IC2006 (a.c. Sra. Célia Maria Carneiro Monteiro). Coordenação de Colaborações Científicas Institucionais. Centro Brasileiro de Pesquisas Físicas. Rua Dr. Sigaud X, 150 Rio de Janeiro, RJ

[14] Beckwith A. How to Use Starobinsky Inflationary Potential Plus Argument From Alder, Bazin, and Schiffer as Radial Acceleration to Obtain First Order Approximation as to Where/when Cosmological Constant May Form. Available from: https://vixra.org/abs/2209.0137

[15] Beckwith A. Creating a (Quantum?) Constraint, in Pre Planckian Space-Time Early Universe via the Einstein Cosmological Constant in a One to One and Onto Comparison between Two Action Integrals. (Text of Talk for FFP 15, Spain November 30, 11 am-11:30 Am, Conference). Available from: http://vixra.org/abs/1711.0355

[16] Beckwith A. A solution of the cosmological constant, using multiverse version of Penrose CCC cosmology, and enhanced quantization compared. Journal of High Energy Physics Gravitation and Cosmology. 2021;7:559-571

[17] Beckwith A. How to use Starobinsky inflationary potential plus argument from Alder, Bazin, and Schiffer as radial acceleration to obtain first order approximation as to where/when cosmological constant may form. Journal of High Energy Physics, Gravitation and Cosmology. 2022;**8**:1012-1018

[18] Linde A. Inflation and Quantum Cosmology. San Diego, California, USA: Academic Press; 1990

[19] Hawking S. The cosmological constant is probably zero. Physics Letters B. 1984;**134**:403-404

[20] Valev D. Neutrino and graviton rest mass estimations by a phenomenological approach. Aerospace Res. Bulg. 2008;**22**: 68-82

[21] Shao L, Wex N, Zhou S. New graviton mass bound from binary pulsars. Physical Review D. 2020;**102**: 024069

[22] Beckwith A. Using Lorentz violation for early universe GW generation due to black hole destruction in the early universe as by freeze. Journal of High Energy Physics, Gravitation and Cosmology. 2021;7:993-1004

[23] Chvanis, P. Self Gravitating Bose-Einstein Condensates", pp. 151-194 'Quantum aspects of black holes", Fundamental Theories of Physics, volume 178, Calmet, X., Editor, by Springer Verlag, Cham, Switzerland, 2014

[24] Shalyt-Margolin AE. Deformed density matrix and quantum entropy of the black hole. Entropy. 2006;**8**:31-43

[25] Shalyt-Margolin A. The density matrix deformation in physics of the early universe and some of its implications. In: Quantum Cosmology Research Trends. Horizons in World Physics. Hauppauge, NY: Nova Science Publishers; 2005;**246**:49-91

[26] Shalyt-Margolin A, Tregubovich A. Deformed density matrix and generalized uncertainty relation in thermodynamics. Modern Physics Letters A. 2004;**19**:71-81

[27] Wesson P. Five Dimensional Physics, Classical and Quantum Consequences of Kaluza–Klein Cosmology. Hackensack, New Jersey, USA: World Scientific Publishing company; 2006

[28] Beckwith A. Examination of a Simpler Version of a Fifth Force Using Variant of Dilaton Model and Padmanabhan Inflaton Scalar Field in Early Universe Conditions. https://vixra.org/abs/2209.0155

[29] Beckwith A. "Refining the Unruh Metric Tensor Uncertainty Principle for a lower bound to Graviton mass as a Compliment to the NLED modification of GR giving an upper bound to a graviton mass". https://vixra.org/pdf/1509.0243v1.pdf

[30] Beckwith A. How 5 Dimensions May Fix a Deterministic Background Spatially as to Be Inserted for HUP in 3+1 Dimensions, and Its Relevance to the Early Universe. Available from: https://vixra.org/abs/2209.0165

[31] Calmet X, Car B, Winstanley E. Quantum black holes. In: Springer Briefs in Physics. New York, USA: Springer New York City; 2014

[32] Beckwith A. Penrose suggestion as to pre Planck era-black holes showing up in

present universe data sets discussed, with a possible candidate as to GW radiation which may provide initial CMBR data. Journal of High Energy Physics, Gravitation and Cosmology. 2021;7:1264-1277

[33] Dolgov A, Ejlli D. Relic gravitational waves from light primordial black holes. Physics Review. 2011;**D84**:024028

Chapter 5

Kantowski-Sachs Barrow Holographic Dark Energy Model in Saez-Ballester Theory of Gravitation

Yendamuri Sobhanbabu, Y. Jnana Prasuna and G. Satyanarayana

Abstract

In this chapter, we have studied a spatially homogeneous and anisotropic Kantowski-Sachs universe in the presence of Barrow Holographic Dark Energy in the background of Saez-Ballester scalar-tensor theory of gravitation. To find the exact solution of the SB field equations, we have assumed that the shear scalar is directly proportional to the expansion scalar. This assumption leads to relation between metric potentials of the models. We have discussed non-interacting and interacting cosmological models. Moreover, we have discussed several cosmological parameters such as energy densities of DM and DE (ρ_m & ρ_b), deceleration (q), equation of state (ω_b) and skewness (α) parameters, squared sound speed (v_s^2), ω_b-$\omega_{b'}$ plane statefinders and Om-diagnostics parameters through graphical representation for both the interacting models. Also, we have observed that the current values of deceleration and EoS parameters of our constructed models coincide with the recent observational data.

Keywords: Kantowski-Sachs, Barrow holographic, scalar-tensor theory, dark energy model, theory of gravitation

1. Introduction

The modern cosmological evidence [1–4] indicated that there is an accelerated expansion. The responsible cause behind this accelerated expansion is a miscellaneous element having exotic negative pressure termed as Dark Energy (DE). The nature and the cosmological origin of DE are still enigmatic. To describe the phenomenon of DE, several models have been presented. According to several findings, DE should behave like a fluid with 'negative pressure, counterbalancing the action of gravity, and speeding up the universe' [5, 6]. The general methodology is to define the dynamics of the universe by assuming the source of DE is represented as a non-zero "cosmological constant Λ," connected to "vacuum quantum field fluctuations" [7, 8]. One proposed solution to DE is the cosmological constant Λ. However, there are difficulties related to its theoretically predicted order of magnitude relative to that of the observed vacuum energy [9]. Other solutions [10–15] go to the idea that cosmic acceleration may be caused by a modification in gravity, perhaps General Relativity (GR) is not valid on cosmological scales.

Hooft [16] has proposed a new dark energy model, known as the Holographic Dark Energy (HDE) model, which was based on the Holographic Principle (HP) and some features of quantum gravity theory. The holographic principle states that the number of degrees of freedom of a gravity-dominated system must vary along with the area of the surface bounding the system [17, 18]. For a system with size L, it is required that the total energy in a region of size L should not exceed the mass of a black hole of the same size for the quantum zero-point energy density associated with the UV cutoff, thus $L^2\rho_b \leq M_p^2$, where ρ_b is the vacuum energy density caused by UV cutoff, M_p is the reduced Planck mass given by the relation $M_p \approx \frac{1}{8\pi G}$ and L is the IR cutoff. The HDE model with Hubble horizon as an IR cutoff is not able to explain the current accelerated expansion [19, 20]. However, the HDE models with other IR cutoffs, e.g., particle horizon, event horizon, apparent horizon, etc. describe the accelerated phenomena of the evolution of the Universe and are in agreement with the observational data [21–27]. Sadri and Khurshudyan [28] have analyzed the HDE model with the Hubble horizon as an IR cutoff in the framework of the flat FRW model while taking into account the non-gravitational interaction between DM and HDE, which is able to explain the current accelerated expansion.

2. Body of the manuscript

Barrow [29, 30] has recently found the possibility that the surface of a black hole could have a complex structure down to arbitrarily tiny due to quantum-gravitational effects. The above potential impacts of the quantum-gravitational space-time form on the horizon region would therefore prompt another black hole entropy relation, the basic concept of black hole thermodynamics. In particular

$$S_B = \left(\frac{A}{A_0}\right)^{1+\frac{\Delta}{2}}, \tag{1}$$

Here A and A_0 stand for the normal horizon area and the Planck area, respectively. The new exponent Δ is the quantum-gravitational deformation with bound as $0 \leq \Delta \leq 1$ [29–33]. The value $\Delta = 1$ gives to the most complex and fractal structure, while $\Delta = 0$ relates to the easiest horizon structure. Here as a special case, the standard Bekenstein-Hawking entropy is re-established and the scenario of Barrow Holographic Dark Energy (BHDE) has been developed. The BHDE models have been explored and discussed by various authors [34–38] in various other contexts. The energy density of BHDE is expressed as

$$\rho_b = CH^{2-\Delta}, \tag{2}$$

where C is an unknown parameter and $\Delta > 0$.

Nandhida and Mathew [39] have considered the Barrow Holographic Dark Energy as a dynamical vacuum, with Granda-Oliveros (GO) length as IR cut-off and studied the evolution of cosmological parameters with the best-estimated model parameters extracted using the combined data-set of supernovae type Ia pantheon (SN-Ia) and observational Hubble's data. Bhardwaj et al. [40] have studied statefinder hierarchy model for the BHDE. Adhikary et al. [41] have constructed a BHDE in the case of non-at universe in particular, considering closed and open spatial geometry and observed that the scenario can describe the thermal history of the universe, with the sequence of matter and DE epochs. Considering BHDE Sarkar and Chattopadhyay [42] reconstruct modified gravity as the form of background evolution and point out that the equation of state can have a transition from quintessence to phantom with the possibility of Little Rip singularity. Saridakis [43] has studied modified cosmology through spacetime thermodynamics and Barrow horizon entropy. Koussour et al. [44, 45] have investigated Bianchi type *I* BHDE model and the stability analysis in symmetric teleparallel gravity.

Shamir and Bhatti [46] have analyzed anisotropic DE Bianchi type *III* cosmological models in Brans-Dicke (BD) theory of gravity. Aditya and Reddy [47] have investigated anisotropic new HDE model in the framework of SB theory of gravitation. Jawad et al. [48] have discussed cosmological implications of Tsallis DE in modified BD theory. Santhi and Sobhanbabu [49, 50] have studied anisotropic THDE models in Scalar tensor theories of gravitation. Sobhanbabu and Santhi [51] have investigated anisotropic MHDE models with sign-changeable interaction in a scalar-tensor theory of gravitation. Recently, Sharif and Majid [52] have studied isotropic and complexity-free deformed solutions in self-interacting in a BD theory of gravitation. Very recently, Pradhan et al. [53] have studied FRW cosmological models with BHDE in the background of scalar-tensor theory of gravitation.

3. Metric and SB field equations

We consider a homogeneous and anisotropic KK Universe described by the line-element

$$ds^2 = dt^2 - X^2(t)dr^2 - Y^2(t)\left(d\theta^2 + \sin^2\theta d\phi^2\right), \tag{3}$$

where $X(t)$ and $Y(t)$ are functions of cosmic time t only.

We assume that the Universe is filled by a DM without pressure with energy density (ρ_m), and BHDE candidate with energy density (ρ_b). Here we take more general Energy Momentum Tensors for DM and BHDE fluid in the following form:

$$T^{\nu}_{\mu} = diag[1,0,0,0]\rho_m \quad and \quad \overline{T}^{\nu}_{\mu} = diag[1, -\omega_b, -(\omega_b + \alpha), -(\omega_b + \alpha)]\rho_b, \tag{4}$$

where $\omega_b = \frac{p_b}{\rho_b}$ is equation of state (EoS) parameter of BHDE, ρ_m is energy density of DM, p_b and ρ_b are pressure and energy density of BHDE, respectively, and α is skewness parameter is in devitation from EoS parameter ω_b on y and z axes, respectively. The SB field equations are

$$R_{\mu\nu} - \frac{1}{2}Rg_{\mu\nu} - w\phi^n\left(\phi_{,\mu}\phi_{,\nu} - \frac{1}{2}g_{\mu\nu}\phi_{,\beta}\phi^{,\beta}\right) = -(T_{\mu\nu} + \overline{T}_{\mu\nu}), \tag{5}$$

where $T_{\mu\nu}$ and $\overline{T}_{\mu\nu}$ are energy-momentum tensors (EMT) for DM and DE, respectively. Scalar field ϕ equation

$$2\phi^n\phi^{,\mu}_{;\mu} + n\phi^{n-1}\phi_{,\beta}\phi^{,\beta} = 0, \tag{6}$$

and energy conservation equations are

$$\left(T_{\mu\nu} + \overline{T}_{\mu\nu}\right)_{;\nu} = 0, \tag{7}$$

where $T_{\mu\nu}$ and $\overline{T}_{\mu\nu}$ are EMTs for DM and BHDE, respectively.

The SB field Eq. (5), for KK line-element Eq.(3) with the help of Eq.(4), can be written as

$$2\frac{\ddot{Y}}{Y} + \frac{\dot{Y}^2}{Y^2} + \frac{1}{Y^2} - \frac{w}{2}\phi^n\dot{\phi}^2 = -\omega_b\,\rho_b, \tag{8}$$

$$\frac{\ddot{X}}{X} + \frac{\ddot{Y}}{Y} + \frac{\dot{X}\dot{Y}}{XY} - \frac{w}{2}\phi^n\dot{\phi}^2 = -(\omega_b + \alpha)\,\rho_b, \tag{9}$$

$$2\frac{\dot{X}\dot{Y}}{XY} + \frac{\dot{Y}^2}{Y^2} + \frac{1}{Y^2} + \frac{w}{2}\phi^n\dot{\phi}^2 = -(\rho_m + \rho_b), \tag{10}$$

$$\ddot{\phi} + \left(\frac{\dot{X}}{X} + 2\frac{\dot{Y}}{Y}\right)\dot{\phi} + \frac{n}{2}\frac{\dot{\phi}^2}{\phi} = 0. \tag{11}$$

We can write the conservation Eq.(7) of the DM and BHDE as

$$\dot{\rho}_m + \left(\frac{\dot{X}}{X} + 2\frac{\dot{Y}}{Y}\right)\rho_m + \dot{\rho}_b + \left(\frac{\dot{X}}{X} + 2\frac{\dot{Y}}{Y}\right)(1 + \omega_b)\rho_b + 2\alpha\left(\frac{\dot{Y}}{Y}\right)\rho_b = 0, \tag{12}$$

where overhead dot (.) denotes ordinary differentiation with respect to cosmic time t.

The SB field eqs. (8)–(11) form a system of four (4) non-linear equations with seven (7) unknowns; X, Y, ρ_m, ρ_b, ω_b, α, and ϕ. In order to solve the field equations explicitly, we need three additional constraints which we will assume in the next section. Now we will know some of the physical and geometric quantities that we will need later.

The average scale parameter of the KK Universe is given by

$$a = \left(XY^2\right)^{\frac{1}{3}}. \tag{13}$$

The spatial volume of the universe

$$V = (a)^3 = XY^2. \tag{14}$$

The average Hubble parameter (H), expansion scalar (θ), and shear scalar (σ^2) of KK universe are defined as

$$H = \frac{\dot{a}}{a},\ \ \theta = 3H,\ \ and\ \sigma^2 = \frac{1}{2}\left[\left(\frac{\dot{X}}{X}\right)^2 + 2\left(\frac{\dot{Y}}{Y}\right)^2\right] - \frac{\theta^2}{6} \tag{15}$$

The Deceleration Parameter (DP) q of the KK universe is defined as

$$q = -\frac{a\ddot{a}}{\dot{a}^2} \tag{16}$$

4. Solution of the field equations and cosmological models

Hence to find the exact solution of the field equations, we have to use some physically viable conditions; The shear scalar (σ^2) is directly proportional to the scalar expansion (θ) which leads to a relationship between metric potentials [54].

$$X = Y^l, \tag{17}$$

where $l \neq 1$ is a positive constant and preserves the non-isotropic behavior of the Universe. Also, we assume that the deceleration parameter (DP) q is a function of the Hubble parameter (H) [55].

$$q = -1 + \gamma H, \tag{18}$$

where γ is an arbitrary constant.
Now using eqs. (16), and (18), we get the exact solution

$$a = e^{\frac{1}{\gamma}\sqrt{2\gamma t + c_1}}, \tag{19}$$

where c_2 is an integration constant and γ arbitrary constant. From Eqs. (13), and (19), we found the metric potentials

$$X = e^{\frac{3l}{\gamma(l+2)}\sqrt{2\gamma t + c_1}}, \quad Y = e^{\frac{3}{\gamma(l+2)}\sqrt{2\gamma t + c_1}} \tag{20}$$

From eq. (2), the energy density of BHDE is

$$\rho_b = C\left[\frac{1}{(2\gamma t + c_1)}\right]^{\frac{2-\Delta}{2}} \tag{21}$$

Thus, the metric corresponding to the metric potentials (20) can be written as

$$ds^2 = dt^2 - e^{\frac{6l}{\gamma(l+2)}\sqrt{2\gamma t + c_1}} dr^2 - e^{\frac{6}{\gamma(l+2)}\sqrt{2\delta t + c_2}}\left(d\theta^2 + \sin^2\theta \, d\phi^2\right) \tag{22}$$

From eqs. (10), (11), (20), and (21), we found the skewness parameter (α) is

$$\alpha = \frac{1}{C}\left[\frac{9(1-l^2)H^{-1} + 3\gamma(l+2)(l-3)}{(l+2)^2 H^{-\frac{3}{2}}} + \frac{9(1-l)}{(l+2)^2 H^{-\frac{1}{2}}} + e^{-\frac{6}{\gamma(l+2)H}}\right] H^{\Delta-2} \tag{23}$$

The scalar field ϕ is

$$\phi = \phi_0\left(\frac{n+2}{2}\right)\int e^{\frac{-3}{\gamma}\sqrt{2\gamma t + c_1}} \, dt, \tag{24}$$

where ϕ_0 is an integration constant.

The plot of DP (q) against redshift (z) is shown in **Figure 1**. We have observed that the DP (q) passes from positive to negative value as the redshift increase and it approaches to -1 at $z = -1$. Thus, our model of the Universe goes from an early deceleration region ($q > 0$) to a current acceleration region ($q > 0$). Also, we have observed that the current values of q is consistent with recent observational data.

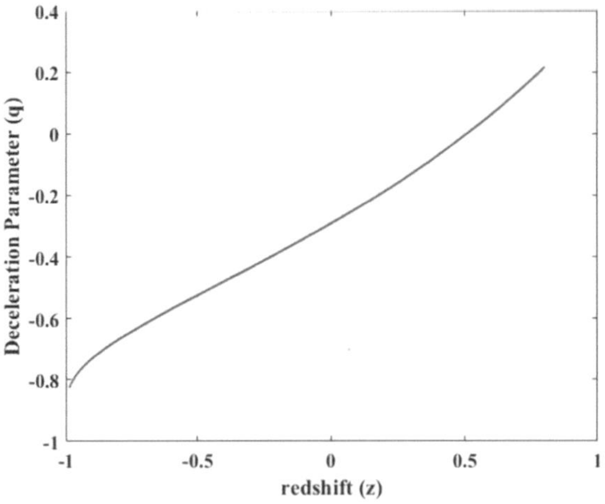

Figure 1.
Variation of deceleration parameter q versus redshift (z) for $l = 1.995$, $c_2 = 1$, $w = 1000$, and $c_3 = 1$ of all models.

4.1 Non-interacting BHDE in the SB cosmology

First, we consider that two fluids (DM and BHDE) do not interact with each other. Hence the conversation eq. (14), of the fluids may be conserved separately. The conservation eq. (14) of barotropic fluid leads to

$$\dot{\rho}_m + 3H\rho_m = 0, \tag{25}$$

whereas the conservation eq. (14) BHDE leads to

$$\dot{\rho}_b + 3H(1 + \omega_b)\rho_b + 2\alpha\rho_b\frac{\dot{Y}}{Y} = 0 \tag{26}$$

From eq. (21) by using eqs. (20), (21), and (23), we get the EoS parameter

$$\omega_b = -1 - \left(\frac{2 - \Delta}{3H^2}\right)\dot{H} - \left(\frac{2}{l + 2}\right)\alpha, \tag{27}$$

where $\dot{H} = -\frac{\gamma}{(\sqrt{2\gamma t + c_1})^{\frac{3}{2}}}$ and $\alpha = \frac{1}{C}\left[\frac{9(1-l^2)H^{-1}+3\gamma(l+2)(l-3)}{(l+2)^2H^{-\frac{3}{2}}} + \frac{9(1-l)}{(l+2)^2H^{-\frac{1}{2}}} + e^{-\frac{6}{\gamma(l+2)H}}\right]H^{\Delta-2}$.

The evolution of DM and BHDE densities with redshift (z) is depicted in **Figures 2** and **3** for various values of Δ, we can see that DM and BHDE densities are positive and increasing functions of redshift z throughout the evolution of the Universe at the present epoch.

In **Figure 4**, we have plotted the behavior of skewness parameter (α) versus redshift (z). It can be seen that the skewness parameter decreasing with an increase in redshift (z) but throughout evolution the skewness parameter (α) is positive.

In **Figure 5**, we observed the dynamics of the EoS parameter (ω_b) against redshift (z) for three various values of $\Delta = 0.92\ 0.94\ 0.96$. The EoS parameter classifies the expansion of the Universe. The EoS parameter ω_b of the BHDE for the non-interacting model completely varies in quintessence region ($-1 < \omega_b < -\frac{1}{3}$). The current values of ω_b are consistent with Planck observational data.

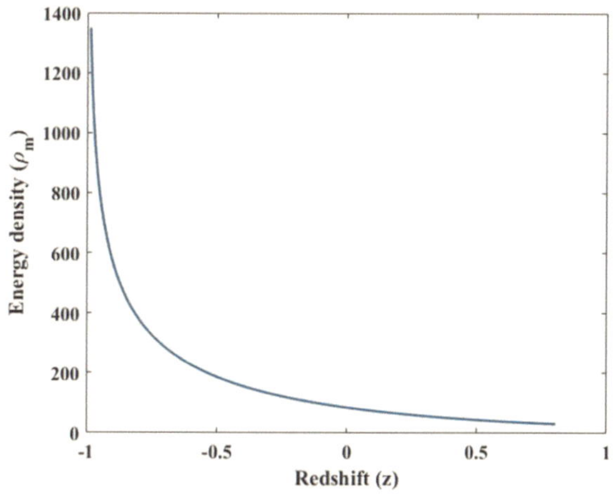

Figure 2.
Variation of energy density (ρ_m) of DM versus redshift (z) for $l = 1.995$, $c_2 = 1$, $w = 1000$, and $c_3 = 1$ of all models.

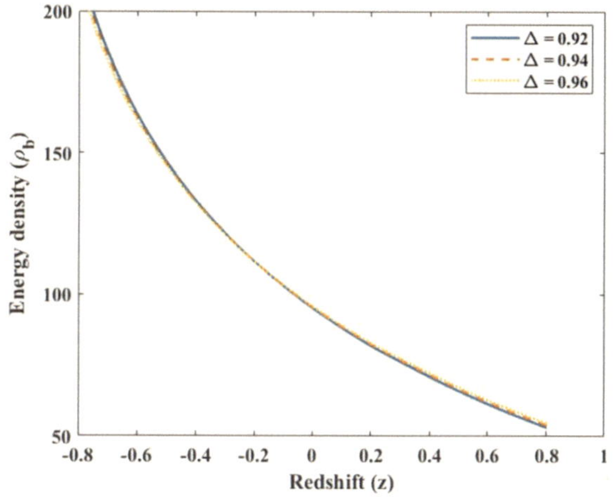

Figure 3.
Variation of energy density (ρ_b) of BHDE versus redshift (z) for.

4.2 $\omega_b - \omega_b{}'$ plane

Caldwell and Linder [56] have pointed out that the quintessence phase of DE can be separated into two distinct regions, that is, thawing ($\omega_{b'} > 0$, $\omega_b < 0$) and freezing ($\omega_{b'} < 0$, $\omega_b < 0$) regions through ω_b-$\omega_{b'}$ plane. Applying the derivative of Eq. (22) with respect to $\ln a$, we have

$$\omega_b{}' = \frac{(\Delta - 2)\left(H\ddot{H} - 2\dot{H}^2\right)}{3H^4} - \frac{2\dot{\alpha}}{(l+2)H}, \tag{28}$$

where $\dot{\alpha} = \frac{1}{C}\left[\frac{1}{(l+2)^2}\left(9(l^2-1)H^{-\frac{1}{2}} + \frac{3}{2}\left(9(1-l^2)H^{-1} + 3\gamma(l+2)(l-3)\right)H^{\frac{1}{2}}\right) + \frac{9(1-l)}{(l+2)^2}H^{-\frac{1}{2}} + \frac{6H}{\gamma(l+2)}e^{\frac{-6}{\gamma(l+2)H^2}} + (\Delta-2)\left(\frac{9(1-l^2)H^{-1}+3\gamma(l+2)(l-3)}{(l+2)^2H^{-\frac{3}{2}}} + e^{-\frac{6}{\gamma(l+2)H}}\right)\right]H^{\Delta-3}\dot{H}$, and $\ddot{H} = \frac{3\gamma^2}{(2\gamma t + c_1)^{\frac{5}{2}}}$.

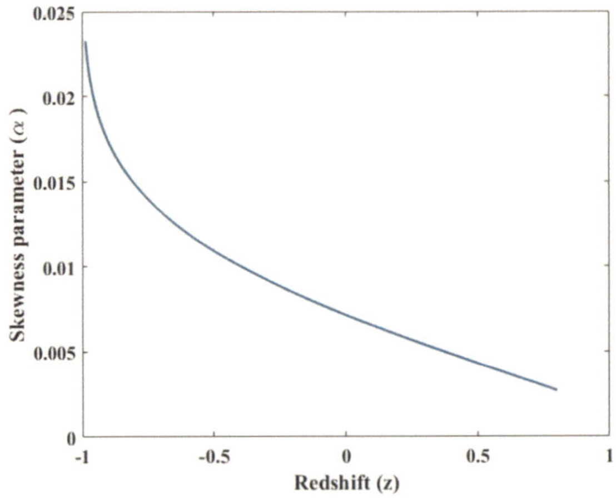

Figure 4.
Variation of skewness parameter (α) versus redshift (z) for.

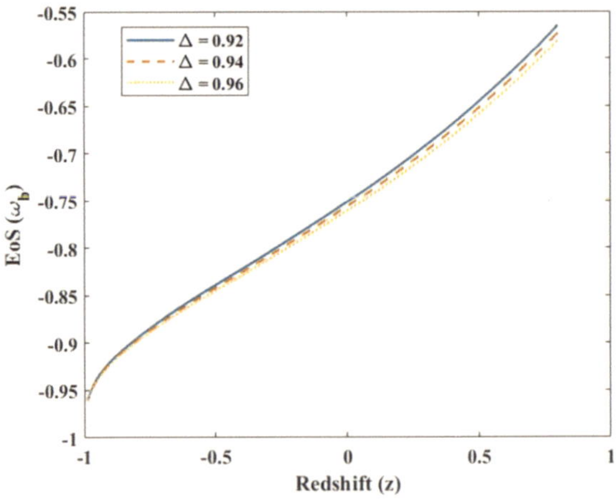

Figure 5.
Variation of equation of state parameter ω_b versus redshift (z) for.

Figure 6 shows the $\omega_b - \omega_{b'}$ plane for the three various values of Δ. We have observed that our non-interacting BHDE model lies in the freezing region ($\omega_b < 0$ and $\omega_{b'} < 0$). It is noticed that the Universe's cosmic expansion accelerates more fastly in this freezing area.

4.3 Stability analysis

We analyze now the stability of the obtained BHDE (non-interacting and interacting) models.

$$v_s^2 = \omega_b + \frac{\rho_b}{\dot{\rho}_b}\, \dot{\omega}_b \tag{29}$$

For our non-interacting BHDE model, squared speed sound v_s^2 is given by

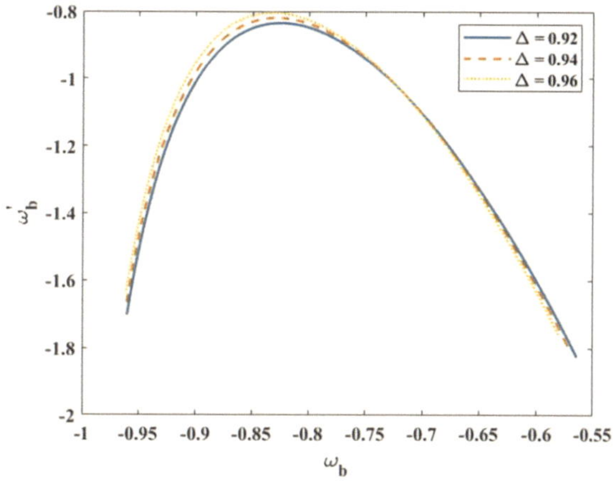

Figure 6.
Variation of ω_b versus $\omega_{b'}$ of the non-interacting model.

$$v_s^2 = -1 - \left(\frac{2-\Delta}{3H^2}\right)\dot{H} - \left(\frac{2}{l+2}\right)\alpha + \left(\frac{H\dot{H}}{2-\Delta}\right)\dot{\omega}_b \tag{30}$$

where $\dot{\omega}_b = \frac{(\Delta-2)\left(H\ddot{H}-2\dot{H}^2\right)}{3H^3} - \frac{2\dot{\alpha}}{l+2}$, *here* $\dot{\alpha} = \frac{1}{C}\left[\frac{1}{(l+2)^2}\left(9(l^2-1)H^{-\frac{1}{2}} + \frac{3}{2}\left(9(1-1)H^{-\frac{1}{2}} + \frac{3}{2}(9(1-l^2)H^{-1}\right.\right.\right.$
$\left.\left.\left. +3\gamma(l+2)(l-3))H^{\frac{1}{2}}\right) + \frac{9(1-l)}{(l+2)^2}H^{-\frac{1}{2}} + \frac{6H}{\gamma(l+2)}e^{\frac{-6}{\gamma(l+2)H^2}} + (\Delta-2)\left(\frac{9\left(1-l^2\right)H^{-1}+3\gamma(l+2)(l-3)}{(l+2)^2H^{-\frac{3}{2}}} + e^{-\frac{6}{\gamma(l+2)H}}\right)\right]H^{\Delta-3}\dot{H}$

For the non-interacting model, **Figure 7** shows the evolution of the SSS in terms of redshift (z). It is clear that the BHDE non-interacting model is initially unstable ($v_s^2 \leq 0$) and with cosmic expansion it becomes stable ($v_s^2 > 0$).

4.4 Interacting BHDE in the SB cosmology

In this case, we focus on the interaction between two dark fluids. Since the nature of both BHDE and DM is still unknown, there is no physical argument to exclude the possible interaction between them. Recently, some observational data shows that

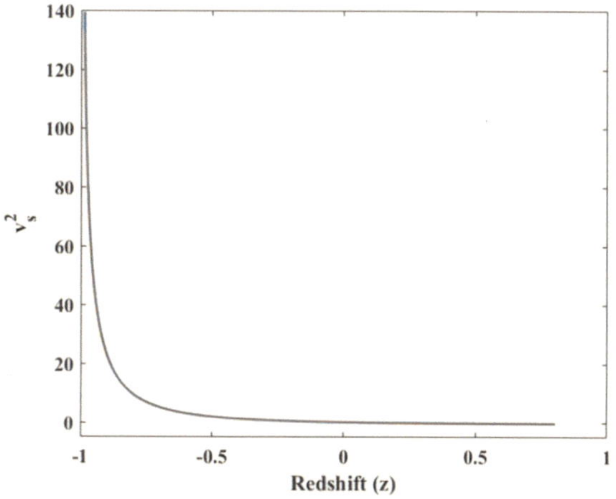

Figure 7.
Variation of squared sound speed v_s^2 versus redshift (z) of the non-interacting model.

there is an interaction between dark sectors [57, 58]. Several authors [59–61] have investigated the signature of interaction between DE and DM by using optical, X-ray, and weak lensing data from the relaxed galaxy clusters. So, it is reasonable to assume the interaction between BHDE and DM in cosmology:

$$\dot{\rho}_m + 3H\rho_m = Q, \tag{31}$$

whereas the conservation eq. (14) BHDE leads to

$$\dot{\rho}_b + 3H(1+\omega_b)\rho_b + 2\alpha\rho_b\frac{\dot{Y}}{Y} = -Q, \tag{32}$$

where Q denotes the interaction term between two fluids (DM and BHDE) and we assume the interaction $Q = 3\beta H\rho_b$, where β is coupling parameter:

Now, from eqs. (20), (21), (23), (24), and (32), we found that the EoS parameter is

$$\omega_b = -1 - \beta - \left(\frac{2-\Delta}{3H^2}\right)\dot{H} - \left(\frac{2}{l+2}\right)\alpha, \tag{33}$$

where $\dot{H} = -\frac{\gamma}{(\sqrt{2\gamma t+c_1})^{\frac{3}{2}}}$ and $\alpha = \frac{1}{C}\left[\frac{9(1-l^2)H^{-1}+3\gamma(l+2)(l-3)}{(l+2)^2H^{-\frac{3}{2}}} + \frac{9(1-l)}{(l+2)^2H^{-\frac{1}{2}}} + e^{-\frac{6}{\gamma(l+2)H}}\right]H^{\Delta-2}$

For interacting BHDE model, the EoS parameter (ω_b) versus redshift (z) for three various values of Δ and β are shown in **Figures 8–10**. We have observed that the EoS parameter starts from the matter-dominated era, then it moves to the quintessence region ($-1 < \omega_b < -\frac{1}{3}$) and crosses the ΛCDM model ($\omega_b = -1$), and finally approaches to phantom region ($\omega_b < -1$). Further, the current values of the EoS parameter (ω_b) are consistent with recent [62] observational data.

Figures 11–13 show the $\omega_b - \omega_{b'}$ plane for the three various values of Δ and β. We have observed that our interacting BHDE model lies in the freezing region ($\omega_b < 0$ and $\omega_{b'} < 0$). It is noticed that the Universe's cosmic expansion accelerates more rapidly in this freezing area.

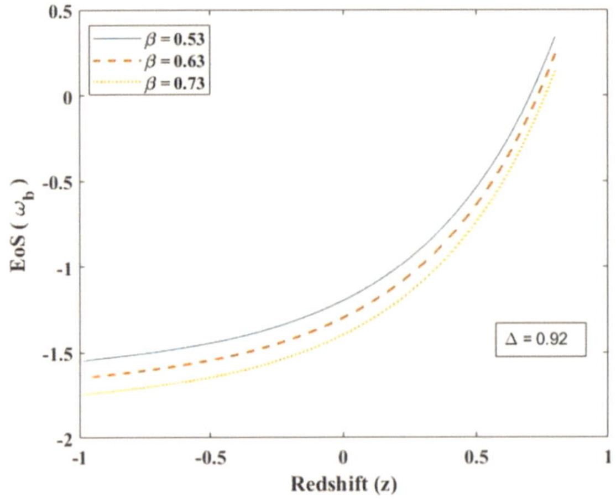

Figure 8.
Variation of ω_b versus redshift (z) for interacting model for $\Delta = 0.92$.

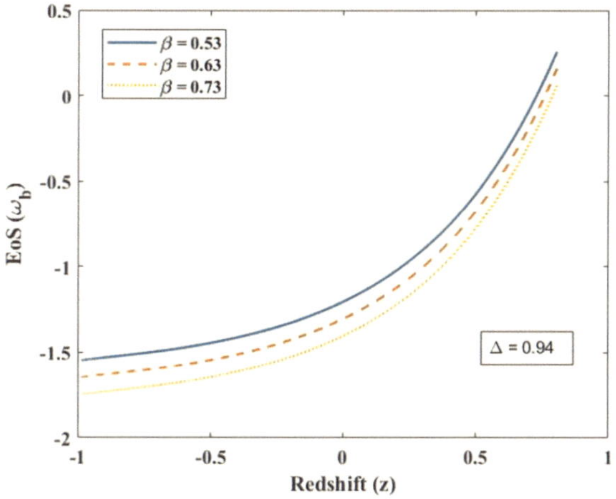

Figure 9.
Variation of ω_b versus redshift (z) for interacting model for $\Delta = 0.94$.

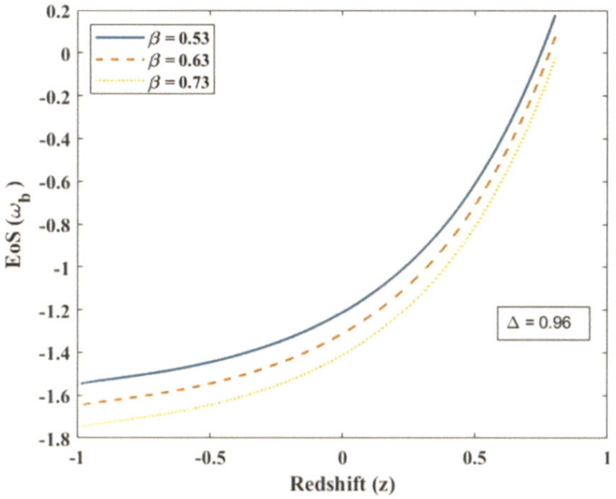

Figure 10.
Variation of ω_b versus redshift (z) for interacting model for $\Delta = 0.96$.

For our interacting BHDE model, **Figure 14** shows the evolution of the SSS in terms of redshift (z). It is clear that the interacting BHDE model is initially unstable $(v_s^2 \leq 0)$ and with cosmic expansion it becomes stable (v_s^2).

4.5 Statefinder diagnostics

In this section, we focus on the diagnosis of the statefinder. The Hubble parameter H represents the Universe's expansion rate and the deceleration parameter q represents the rate of acceleration or deceleration of the expanding cosmos, which are two well-known geometrical parameters that characterize the Universe's expansion history. They only depend on the average scale parameter a. This statefinder pairs {r,s} [63, 64] as follows

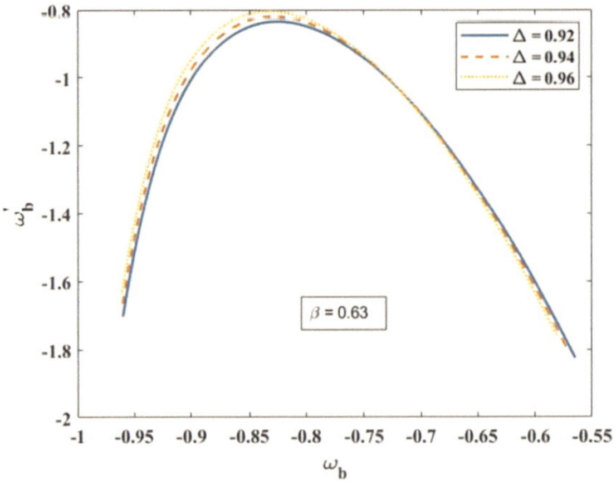

Figure 11.
Variation of ω_b versus $\omega_{b'}$ of the interacting model.

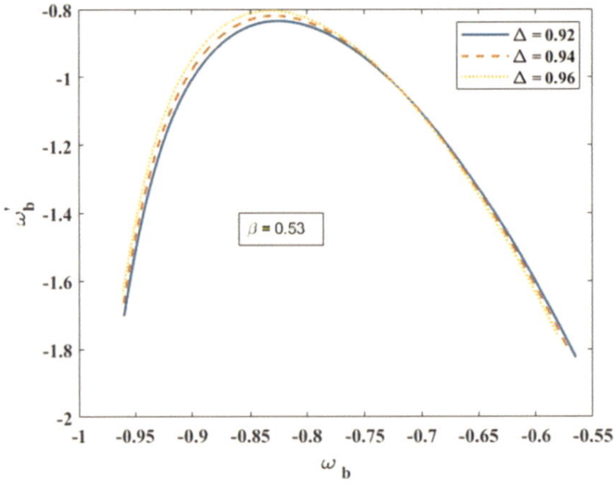

Figure 12.
Variation of ω_b versus $\omega_{b'}$ of the interacting model.

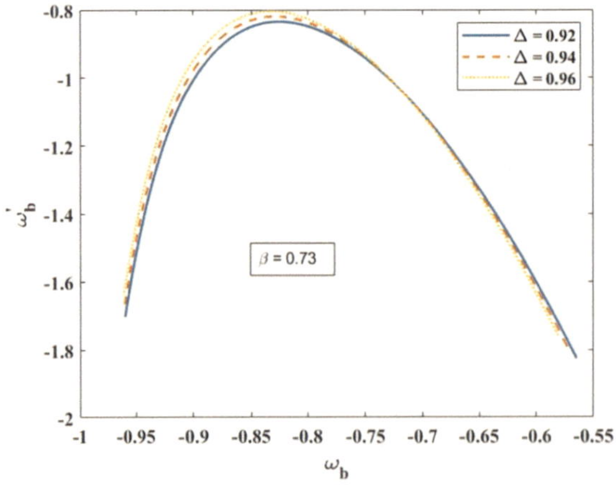

Figure 13.
Variation of ω_b versus $\omega_{b'}$ of the interacting model.

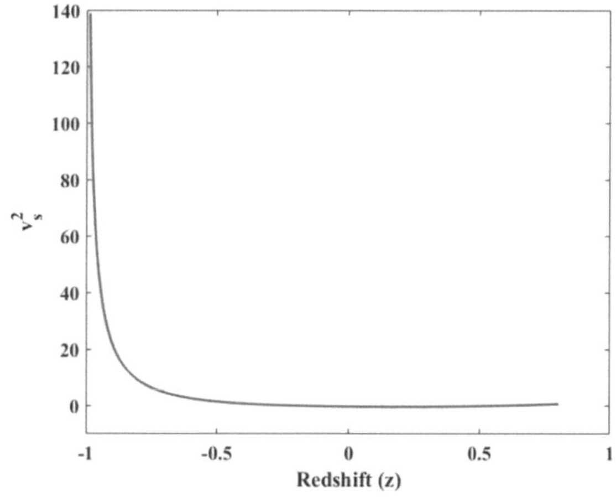

Figure 14.
Variation of squared sound speed (v_s^2) versus redshift (z) of the interacting models.

$$r = \frac{\ddot{a}}{aH^3} = \frac{\ddot{H}}{H^3} + 3\frac{\dot{H}}{H^2} + 1 \tag{34}$$

$$s = \frac{r-1}{3\left(q-\frac{1}{2}\right)} = \frac{\frac{\ddot{H}}{H^3} + 3\frac{\dot{H}}{H^2}}{3\left(q-\frac{1}{2}\right)} \tag{35}$$

Figure 15 shows the evolutionary trajectories in $\{r, s\}$ plane. In the figure, our constructed model lies in the chaplygin gas ($\{r > 1, s < 0\}$) model and also meets ΛCDM model ($\{r = 11, s = 0\}$). **Figure 16** depicts that our model lies in Standard Cold Dark Matter (SCDM) region ($\{r > 1, q > 0\}$)and also meets ΛCDM region ($\{r = 1, q = 0\}$).

4.6 Om-diagnostic

As a complementary to the statefinder parameters $\{r, s\}$, a new diagnostic is known as *Om* studied by some of the researchers [65, 66]. The *Om* diagnostic parameter for our model is

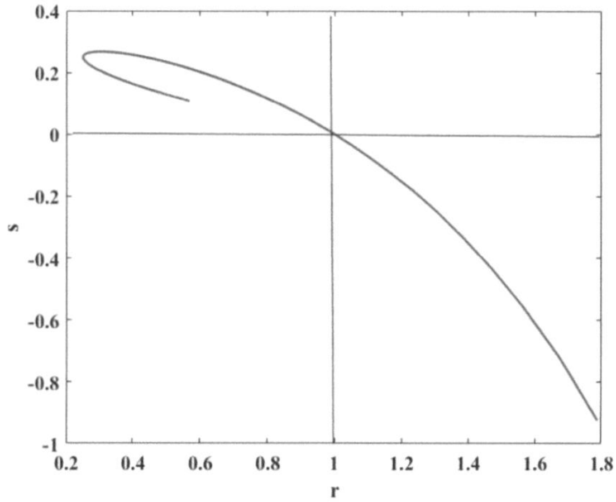

Figure 15.
Variation of statefinder parameters r versus s of the models.

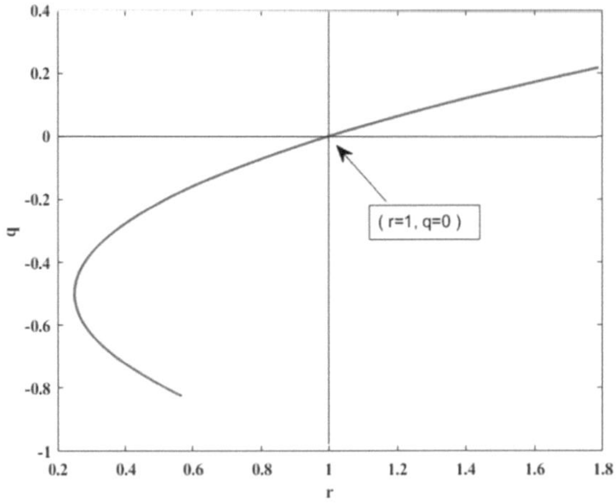

Figure 16.
Variation of statefinder parameters r versus s of the models.

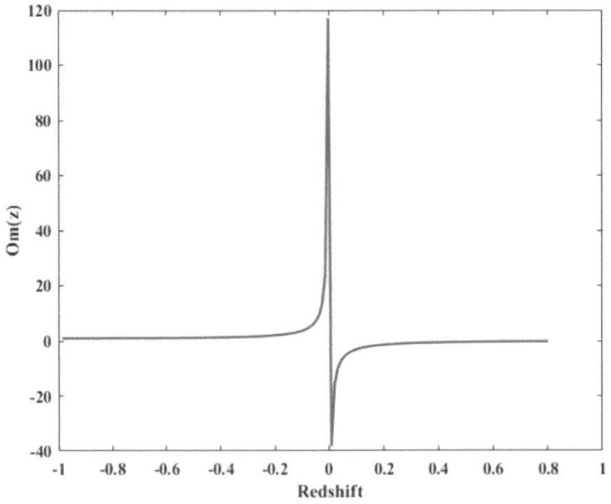

Figure 17.
Variation of statefinder parameters r versus s of the models.

$$Om(x) = \frac{h^2(x) - 1}{x^3 - 1}, \tag{36}$$

where $x = 1 + z$ and $h(x) = \frac{H(x)}{H_0}$.

The trajectory of Om diagnostics versus redshift (z) is shown in **Figure 17**. The trajectory reveals that the BHDE model shows initially a positive slope of the trajectory indicating that our model has phantom behavior and the negative slope of the trajectory indicates that our model behavior is quintessence in late time.

5. Conclusions

In this chapter, we have investigated the accelerated expansion by assuming the BHDE Universe within the framework of SB scalar-tensor theory of gravity. We have investigated various cosmological parameters to analyze the viability of the models and our conclusions are the following:

The deceleration parameter (q) passes from positive to negative value as the redshift increase and it approaches to -1 at $z = -1$. Thus, our model of the Universe goes from an early deceleration region $(q > 0)$ to a current acceleration region (**Figure 1**). The parameter (q) of our model consistent with the current observational data are, Capozziello et al. [67], given as

$$q = -0.6401 \pm 0.187(BAO + Masers + TDSL + Panthelon + H_0)$$
$$q = -0.930 \pm 0.218(BAO + Masers + TDSL + Panthelon + H_z).$$

For our non-interacting model, the energy densities of DM and BHDE are positive and increasing function of redshift z throughout the evolution of the Universe at the present epoch (**Figures 2** and **3**). The skewness parameter decreasing with increase in redshift (z) but throughout evolution the skewness parameter (α) is positive (**Figure 4**). The EoS parameter ω_b of the BHDE for non-interacting model completely varies in quintessence region $(-1 < \omega_b < -\frac{1}{3})$ for three different values of $\Delta = 0.92 \;\; 0.94 \;\; 0.96$. The current values of ω_b are consistent with Planck [62] observational data (**Figure 5**). The $\omega_b - \omega_{b'}$ plane for the three various values of Δ we observe that our non-interacting BHDE model lies in the freezing region ($\omega_b < 0$ and $\omega_{b'} < 0$). It is noticed that the Universe‘s cosmic expansion accelerates more fastly in this freezing area (**Figure 6**). The SSS is initially unstable $(v_s^2 \leq 0)$ and with cosmic expansion, it becomes stable (v_s^2) for non-interacting BHDE model (**Figure 7**).

For interacting BHDE model, the EoS parameter starts from the matter dominated era, then it moves to the quintessence region $(-1 < \omega_b < -\frac{1}{3})$ and crosses the ΛCDM model $(\omega_b = -1)$, and finally approaches to phatom region $(\omega_b < -1)$ for three different values of Δ and β. Also, it is worthwhile to mention here that the present values of the EoS parameter of our BHDE models are in agreement with the modern Plank observational data given by Aghanim et al. [62]. It gives the constraints on the EoS parameter of dark energy as follows:

$$\omega_b = -1.56^{+0.60}_{-0.48}(Planck + TT + lowE)$$
$$\omega_b = -1.58^{+0.52}_{-0.41}(Planck + TT, EE + lowE)$$
$$\omega_b = -1.57^{+0.50}_{-0.40}(Planck + TT, TE, EE + lowE + lensing)$$
$$\omega_b = -1.04^{+0.10}_{-0.10}(Planck + TT, TE, EE + lowE + lensing + BAO)$$

It can be observed from **Figures 5, 8–10** that the EoS parameter of our models in both non-interacting and interacting cases lie within the above observational limits which shows the consistency of our results with the above cosmological data. We have observed that our interacting BHDE model lies in the freezing region ($\omega_b < 0$ and $\omega_{b'} < 0$) for the three various values of Δ and β. It is noticed that the Universe‘s cosmic expansion accelerates more rapidly in this freezing area (**Figures 11–13**). The SSS is initially unstable $(v_s^2 \leq 0)$ and with cosmic expansion it becomes stable (v_s^2). It is exactly similar to the non-interacting case (**Figur 14**).

The behavior of $\{r, s\}$ and $\{r, q\}$ planes for our model lies in the chaplygin gas $(\{r > 1, s < 0\})$ model and meets ΛCDM model $(\{r = 11, s = 0\})$. The $\{r, q\}$ plane lies in SCDM region $(\{r > 1, q > 0\})$and also meets ΛCDM region $(\{r = 1, q = 0\})$. The trajectory of Om-diagnostics reveals that the BHDE model shows initially our model has phantom behavior and quintessence behavior in late time (**Figures 15–17**).

Finally, we can state that some of the preceding conclusions in KK BHDE model are good agreement with recent observations.

Conflict of interest

The authors declare that they have no known competing financial interests or personal relationships that could have appeared to influence the work reported in this chapter.

Data availability statement

This chapter has no associated data or the data will not be deposited.

Author details

Yendamuri Sobhanbabu[1,2*†], Y. Jnana Prasuna[2†] and G. Satyanarayana[3†]

1 SRKR Engineering College (A), Bhimavaram, India

2 KGRL PG Course, Bhimavaram, India

3 Sasi Institute of Technology and Engineering College, Tadepalligudem, India

*Address all correspondence to: sobhan.maths@gmail.com

† These authors contributed equally.

References

[1] Riess AG et al. Observational Evidence from Supernovae for an Accelerating Universe and a Cosmological Constant. Astronomy Journal. 1998;**116**:1009

[2] Perlmutter S et al. Measurements of Ω and Λ from 42 High-Redshift Supernovae. The Astrophysical Journal. 1999;**517**:565

[3] Spergel DN et al. First-Year Wilkinson Microwave Anisotropy Probe (WMAP)* Observations: Determination of Cosmological Parameters. Astrophysical Journal of Supplement. 2003;**148**:175

[4] Tegmark M et al. Cosmological parameters from SDSS and WMAP. Physics Review D. 2004;**69**:103501

[5] Copeland EJ et al. Dynamics of dark energy. International Journal of Modern Physics D: Gravitation; Astrophysics and Cosmology. 2006;**15**:1753

[6] Ade PAR et al. Cosmological parameters. Astronomy and Astrophysics. 2016;**594**:A13

[7] Perlmutter S et al. Discovery of a supernova explosion at half the age of the Universe. Nature. 1998;**391**:51

[8] Schmidt BP et al. The high-z supernova search: measuring cosmic deceleration and global curvature of the universe using type Ia supernovae. The Astrophysical Journal. 1998;**507**:46

[9] Riess AG et al. Type ia supernova discoveries at z > 1 from the hubble space telescope: Evidence for past deceleration and constraints on dark energy evolution. The Astrophysical Journal. 2004;**607**:665

[10] Barreiro T et al. Quintessence arising from exponential potentials. Physics Review D. 2000;**61**:127301

[11] Caldwell et al. Phantom Energy: Dark Energy with $\omega < -1$: Causes a Cosmic Doomsday. Physical Review Letters. 2003;**91**:071301

[12] Bagla JS et al. Cosmology with tachyon field as dark energy. Physics Review D. 2003;**67**:063504

[13] Armendariz-Picon et al. Essentials of k-essence. Physics Review D. 2001;**63**: 103510

[14] Cognola G et al. Dark energy in modified Gauss-Bonnet gravity: Late-time acceleration and the hierarchy problem. Physics Review D. 2006;**73**: 084007

[15] Ferraro et al. Modified teleparallel gravity: Inflation without an inflaton. Physics Review D. 2007;**75**:084031

[16] 't Hooft G. Dimensional Reduction in Quantum Gravity. 1993, arXiv e-prints, pp gr-qc/9310026

[17] Susskind L. The World as a Hologram. Journal of Mathematical Physics. 1995;**36**:6377

[18] Cohen A et al. Effective Field Theory, Black Holes, and the Cosmological Constant. Physical Review Letters. 1999;**82**:4971

[19] Hsu SD. Holographic Bound in Quantum Field Energy Density and Cosmological Constant. Physics Letters B. 2004;**594**:13

[20] Li M. The Interacting Generalized Ricci Dark Energy Model in Non-Flat Universe. Physics Letters B. 2004;**603**:1

[21] Zhang X, Wu FQ. Constraints on holographic dark energy from type Ia supernova observations. Physics Review D. 2005;**72**:043524

[22] Feng C et al. Journal of Cosmology and Astroparticle Physics. 2007;**09**:005

[23] Li M et al. Holographic dark energy models: a comparison from the latest observational data:. Journal of Cosmology and Astroparticle Physics. 2009;**06**:036

[24] Luongo O. A Thermodynamic Approach to Holographic Dark Energy. Advances High Energy Physics. 2017;**1**: 1424503. DOI: 10.1155/2017/1424503

[25] Setare MR. Interacting Dark Fluids in LRS Bianchi Type-II Universe. Physics Letters B. 2007;**654**:1

[26] Saridakis EN. Physics Letters B. 2008;**661**:335

[27] Singh CP, Srivastava M. European Physical Journal C: Particles and Fields. 2018;**78**:190

[28] Sadri E, Khurshudyan M. On An Interacting New Holographic Dark Energy Model: Observational constraints. International Journal of Modern Physics D: Gravitation; Astrophysics and Cosmology. 2019;**28**:1950152

[29] Saridakis EN. Barrow holographic dark energy: Physics Review D. 2000; **102**123525

[30] Pradhan A et al. A new class of holographic dark energy models in LRS Bianchi Type-I. International Journal of Modern Physics A. 2021;**36**:4

[31] Barrow JD. Physics Letters B. 2020; **808**:135643

[32] Saridakis SN, Basilakas S. The generalized second law of thermodynamics with Barrow entropy. The European Physical Journal C: 2000; **81**:644

[33] Anagnostopoulos FK et al. 2020. DOI: 10.48550/arXiv.2005.10302

[34] Barrow JD et al. Big Bang Nucleosynthesis constraints on Barrow entropy. Physics Letters B. 2021:**136134**

[35] Srivastava S, Sharma UK. Barrow holographic dark energy with Hubble horizon as IR cutoff. International Journal of Geometric Methods in Modern Physics. 2021;**18**:2150014.

[36] Mamon AA et al. Dynamics of an interacting barrow holographic dark energy model and its thermodynamic implications. The European Physical Journal Plus. 2021;**136**134

[37] Abreu EMC, Neto JA. Barrow black hole corrected-entropy model and Tsallis nonextensivity. Physics Letters B. 2020; **810**:135805

[38] Abreu EMC, Neto JA. Barrow fractal entropy and the black hole quasinormal modes. Physics Letters B. 2020;**807**:135602

[39] Nandhida PK, Mathew TK. Barrow holographic dark energy model with GO cut-off – An alternative perspective. International Journal of Modern Physics D. 2022;**31**:2250107

[40] Bhardwaj VK et al. Statefinder hierarchy model for the Barrow holographic dark energy. New Astronomy. 2021;**88**:101623

[41] Dudas E, et al. Slow and safe gravitinos. Physics Review D. 2021;**104**: 123519

[42] Sarkar A, Chattopadhyay S. The barrow holographic dark energy-based

reconstruction off (R) gravity and cosmology with Nojiri–Odintsov cutoff. International Journal of Geometric Methods in Modern Physics. 2021;**18**: 2150148

[43] Saridakis EN. Modified cosmology through spacetime thermodynamics and Barrow horizon entropy. Journal of Cosmology and Astroparticle Physics. 2020;**07**:031

[44] Koussour M et al. Bianchi type-I Barrow holographic dark energy model in symmetric teleparallel gravity. International Journal of Modern Physics A. 2022;**37**:2250184

[45] Koussour M, Bennai M. Stability analysis of anisotropic Bianchi type-I cosmological model in teleparallel gravity. Classical and Quantum Gravity. 2022;**39**: 105001

[46] Shamir MF, Bhatti AA. Canadian Journal of Physics. 2012;**90**:2

[47] Aditya Y, Reddy DRK. Anisotropic new holographic dark energy model in Saez–Ballester theory of gravitation. Astrophysics and Space Science. 2018;**363**: 207

[48] Jawad A et al. Cosmological consequences and thermodynamics of modified gravity with extended nonminimal derivative couplings. International Journal of Modern Physics D. 2019;**28**:1950146

[49] Vijaya Santhi M, Sobhan Babu Y. Bianchi type-III Tsallis holographic dark energy model in Saez–Ballester theory of gravitation. European Physical Journal C. 2020;**80**:1198

[50] Vijaya Santhi M, Sobhan Babu Y. Tsallis holographic dark energy models in Bianchi type space time. New Astronomy. 2021;**89**:101648

[51] Sobhanbabu Y, Vijaya Santhi M. Kantowski–Sachs Tsallis holographic dark energy model with sign-changeable interaction. The European Physical Journal C. 2021;**81**:1040

[52] Sharif M, Majid A. Isotropic and complexity-free deformed solutions in self-interacting Brans–Dicke gravity. International Journal of Modern Physics. 2022;**31**:2240003

[53] Pradhan A et al. FRW cosmological models with Barrow holographic dark energy in Brans–Dicke theory. International Journal of Geometric Methods in Modern Physics. 2022;**19**: 2250106

[54] Collins B et al. Exact Spatially Homogeneous Cosmologies. General Relativity and Gravitation. 1980;**12**:805

[55] Tiwari RK et al. Cosmological models with viscous fluid and variable deceleration parameter. European Physical Journal Plus. 2017;**132**:1

[56] Caldwell RR, Linder EV. The Limits of Quintessence. Physical Review Letters. 2005;**95**:141301

[57] Bertolami O et al. Dark energy–dark matter interaction and putative violation of the equivalence principle from the Abell cluster A586. Physics Letters B. 2007;**654**:165

[58] Bertolami O et al. The Abell cluster A586 and the detection of violation of the equivalence principle. General Relativity and Gravitation. 2009;**41**:2839

[59] Abdalla E et al. Signature of the interaction between dark energy and dark matter in galaxy clusters. Physics Letters B. 2009;**673**:107

[60] Abdalla E et al. Astrophysical uncertainties of dark matter direct

detection experiments. Physical Review D. 2010;**82**:023508

[61] Sobhanbabu Y, Vijaya Santhi M. Anisotropic new agegraphic dark energy model with sign-changeable interaction in a scalar-tensor theory. The European Physical Jounal Plus. 2022;**137**:753

[62] Alam U et al. Exploring the Expanding Universe and Dark Energy using the Statefinder Diagnostic. Monthly Notices of the Royal Astronomical Society. 2003;**344**:1057

[63] Sahni V et al. Statefinder—A new geometrical diagnostic of dark energy. JETP Letters. 2003;**77**:201

[64] Aghanim N et al. Planck 2018 results. VI. Cosmological parameters. arXiv:1807.06209v2. 2018

[65] Zunckel C. Consistency Tests for the Cosmological Constant. Physical Review Letters. 2008;**101**:181301

[66] Shahalam M et al. Om diagnostic applied to scalar field models and slowing down of cosmic acceleration. Monthly Notices of the Royal Astronomical Society. 2015;**448**:2948

[67] Capozziello S. Model-independent constraints on dark energy evolution from low-redshift observations. Monthly Notices of the Royal Astronomical Society. 2019;**484**:4484

Section 2

Observation

Chapter 6

Main Experiments for Detection of Gravitational Waves at Frequency below 3 kHz: A Quick Review

Carlos Frajuca

Abstract

Gravitational Waves were detected at last with a laser-interferometric detector in 2015 with a 4 kilometers long laser-interferometric detector. It took more than 100 years of effort to reach such a goal. This achievement is one more piece to prove the Einstein General Theory of Relativity. Besides new detections with these experiments, a lot of effort has been allocated to the current laser-interferometric detector to improve its performance and detect signals from sources farther away with the intention of searching all the known Universe for Gravitational Wave sources. Nevertheless, this kind of experiment has a frequency range limited by seismic noise around 10 Hz and lower. Efforts are being made for the detection of Gravitational Waves at different frequencies, for instance, laser interferometer in space, measurement of pulsar timing and deviations of polarization of the microwave background. All these experiments are discussed in this chapter as their sources. A very broad frequency range of detectors should be available in the next decade.

Keywords: gravitational waves, gravitational waves detection, pulsar time array, pulsar time, laser interferometer

1. Introduction

Gravitational waves (GW) are space–time ripples generated by accelerated massive objects, to have a reasonable intensity, these massive objects must be of a cosmic origin like compact stars such as Neutron Stars or Black Holes. Considering the Einstein General Theory of Relativity, these GW move at the speed of light and can accelerate masses or excite quadrupolar normal-modes of elastic bodies as the equivalent principle predicts. The potential sources of GWs include binary compact star systems composed of white dwarfs, neutron stars and black holes.

The existence of GW is also a consequence of the Lorentz Covariance of Einstein General Theory of Relativity. GW does not exist in the Newtonian theory of gravitation, which postulates that physical interactions propagate at infinite speed.

The first detection of GWs came from the inward spiral and merger of the Black Hole (BH) binary. The event is called GW150914 [1], and the name is given by the Letters GW followed by the year, month and day of the detection, then the detection happened on September 14, 2015. In 2017 a simultaneous detection of GW radiation and electromagnetic radiation was made from what is called a hipernova, a binary Neutron Star (NS) system that merged. It is called GW170817, which could have started the era of multi-messenger astronomy [2], which involves GW, optical radiation, radio, gamma-rays and X-rays radiation. Studying the universe with these different types of radiation opens a new era for understanding the universe.

In 2017, the Nobel Prize in Physics was awarded to Barry Barish, Kipe Thorne and Rainer Weiss for their role in detecting gravitational waves.

Before the first GW detection, there was indirect evidence of these GW. Measurements of the Hulse and Taylor binary pulsar system suggested that GW was more than a hypothetical concept, at least for the emission of such waves; the authors of this measurement won the Nobel prize. This system is one of the potential sources of detectable gravitational waves, but because of the frequency with which this system operates, a new kind of detector can detect much smaller frequencies than the one that detected GW150914. The kind of detector responsible for this detection are interferometric ones, whose sensitivity is limited at frequencies close to 10 Hz and lower because of seismic noise on the mirrors.

Another option to detect GW radiation is using resonant-mass detectors (this was the first kind of detector proposed in the 1960s). From those, the only remaining detector is the Mario Schenberg Brazilian GW detector that uses the detection of the vibration modes of five quadrupole modes of a spherical resonant-mass of 1124 kg with a radius of 32.33 cm made of CuAl6% alloy [3], this mass vibrates when a GW passes through it with a resonant frequency. **Figure 1** shows a schematic of such a detector that operates at a temperature of 4 K.

Figure 1.
Drawing perspective of Schenberg detector.

Detectors for GW are the topic of this book chapter and will be addressed in the next section. Then, in the following section, a spectrum with the main sources and main experiments will be shown.

2. Gravitational waves experiments

2.1 Laser-interferometric detectors

The Interferometry system essentially works by measuring the variations that occur in laser light beams, which are arranged along two orthogonal arms, as a normal interferometer works, the main difference is the size of these detectors, with arm lengths around 3–4 kilometers long. The detection occurs when the variations in the interferometer arm's length cause variations in the interference patterns in the photo-detectors that are observed because the velocity of the light beams is constant even in the presence of GW. This difference could also cause a difference of arriving time in the mirrors, but the time difference is too small to be measured. Using arm length of kilometer size makes these detectors capable of measurements in length of less than a thousandth of a neutron diameter.

A powerful laser beam (with a power of tens of Watts) passes through a beam splitter allowing the two generated beams to have the same phase and to be separated orthogonally, at the end they are reflected by mirrors that exist in the ends of the arms, some are semi-reflective (the power in each laser beam inside the arms reach 400 kW). The phases of the reflected laser beams are adjusted to generate a destructive interference pattern at the photodetector, so no signal is detected by the photodetector in the detector with no GW passing through. For the occasion of a GW across it, it causes space–time to expand and contract infinitesimally in orthogonal directions, thus changing the interference pattern in the photodetector and a signal is detected. **Figure 2** shows the schematics of such a detector.

There are two such detectors in operation: the Laser Interferometric Gravitational Observatory (LIGO) detector (an American detector [4]) and VIRGO (a Franco-Italian detector [5]). A third detector is being built in Japan, the Kagra detector, as can be seen in [6]. Efforts are being made to build a fourth detector to complete the sky coverage for GW in the range of 10–3000 Hz; the position of this next detector is probably in India.

Also, the existence of more than one of these detectors is a way of preventing the possibility of false detections due to small earthquakes, vibrations in the mirror

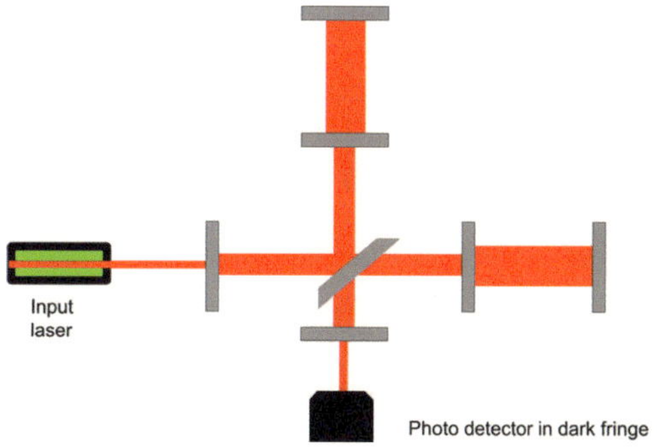

Figure 2.
Schematics of a laser-interferometric GW detector.

suspensions, some phase variations, or some other local source of noise. When detecting a signal, this signal will be compared with signals detected by other detector or detectors, and the signal is only confirmed if signals are measured in more than one detector, with a right coincidence time between the detectors and the signals have the same characteristics: exactly the same profile in frequencies considering the detector characteristics and the arriving time is coherent with the position of the detectors. All of these measurements happen in a vacuum, avoiding the possibility of interactions of the laser in some molecules that could mimic the effect of a GW signal. The characterization of one of these detectors can be seen in [1]. Some of these are: lasers operating in higher frequency, mirror suspension made of fused silica to reduce creep noise and the improvement of mirror suspension to reduce thermal and seismic noise.

2.1.1 The real laser-interferometric detector

The interferometric GW detectors are very complicated and complex machines. As the interferometer must be set in a dark fringe condition, the vibrations acting on the mirrors can change the dark fringe condition as the mirrors are somehow connected to the ground and are finite temperature, so they vibrate, which changes the dark fringe condition. Then a very good suspension that attenuates the vibrations must be used, even adding some active system to lower the vibration. **Figure 3** shows a schematics of such suspension; this example is about the LIGO detector.

The Italian detector Virgo has a more sophisticated suspension, which makes this detector more sensitive at lower frequencies, for each mirror suspension is composed of an inverted pendulum and six masses suspended by their centers. As it is not enough, add to it a collection of 18 LVDTs (Linear Variable Displacement Transducers), five accelerometers, 23 coils, three piezoelectric devices and 21 motor drivers for each mirror. All this system is called the supperatenuator [5].

These details show that it is very difficult to keep the GW interferometric detectors locked in a dark fringe condition, even depending on active systems.

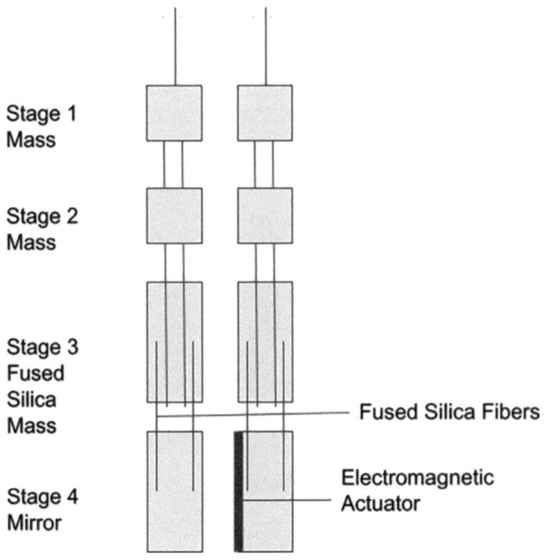

Figure 3.
Schematics side view of the mirror suspension system of the LIGO detector showing the electrostatic actuator which is used to keep the detector locked in at a dark fringe condition.

But how sensitive the interferometer must be to make measurements of GW. For a 4 kilometer size GW detector, the first measurement of GW was of a displacement of 10^{-18}. As the arm length is 4 kilometer, the variation in length was 4×10^{-15} m. As the power inside the arms was 100 kW with an input power of 20 W, the detector has a recycling factor of 5000, which makes the real sensitivity of the interferometer close to 10^{-12} m.

2.2 The resonant-mass gravitational wave detector

Figure 4 shows an example of a second-generation resonant-mass GW detector which operated for about two decades at Louisiana State University. This detector was called Allegro, as can be seen in [7]. This kind of detector operating in their quantum limit can be used to calibrate interferometric GW detectors, as can be seen in [8, 9], as these detectors are narrowband they cannot give the behavior of the GW with the frequency.

When a GW passes through a resonant-mass GW detector, its resonant-mass (called antenna) vibrates in resonance with the GW. Then, the antenna surface motion is measured by motion sensors which are transducers that transform these vibrations into electrical signals. These signals are then analyzed and the intensity of such GW can be obtained by modeling the system detection. The transducer is usually composed of a mechanical oscillators that increase the coupling of the vibration modes of the antenna to the electronic sensor, selecting and filtering the frequency of interest [10–13] and (for a spherical detector) the direction of the signal can be

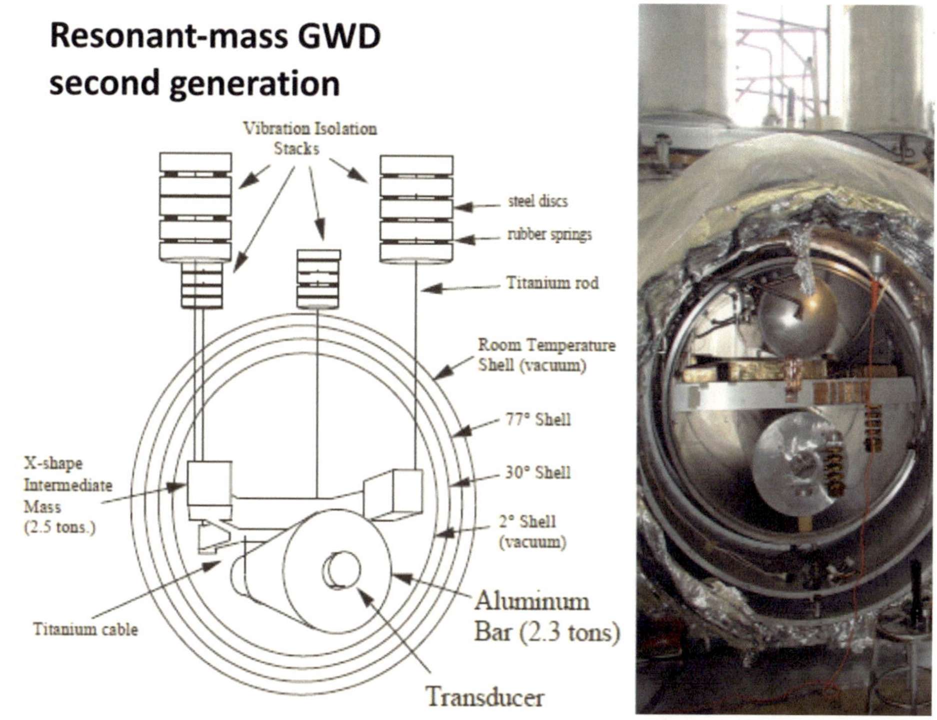

Figure 4.
The figure shows the schematics of a bar resonant-mass gravitational wave detector, in this case, the detector called Allegro, which operated for two decades in the Louisiana State University.

obtained, as can be seen in [14–16]. The Brazilian detector uses microwave parametric transducers that match the antenna's 3.2 kHz quadrupolar mode to the electronic sensor. This transducer has a superconducting cavity into which a resonant, very-low-noise monochromatic microwave signal is injected, and changes in this signal as the cavity vibrates, allowing the measurement of the GW's effects on the antenna. Other options are capacitive and inductive transducers.

When no GW is present in the system, the transducer is never still, as it vibrates because of the thermal noise, the noise that usually limits the sensitivity of a resonant-mass GW detector.

When GW resonant-mass detectors operated for about two decades and formed a worldwide network, this network set upper limits for amplitudes of GW signals in their operational frequency range that they covered while refining its sensitivity. **Figure 5** shows this worldwide network (when its third-generation antennas were operational). Those antennas were cylindrical bars tuned to GW with frequencies around 1 kHz.

The design of resonant-mass GW detectors did not allow their antennas to vibrate frequencies less than 1 kHz [17]. After the report of direct detections and the following direct detections, the frequency range around 128 Hz seems to be a possible range for GW detection. One of the reasons these detectors never made a detection was the operational frequency chosen because there is no signal in this frequency region. A possibility to use these detectors would be to change their design and develop a new resonant-mass GW detector with quadrupole modes near the frequency of 128 kHz that could operate in coincidence with laser-interferometric detectors.

These detectors are easier and cheaper to build. Such massive detectors properly designed and positioned near interferometric GW detectors can be used to make coincidence detections of the stochastic GW background, as it can be used to provide data for veto procedures (which help identify false detections), helping more accuracy in the data analysis. They can also be used to increase calibration, as mentioned before.

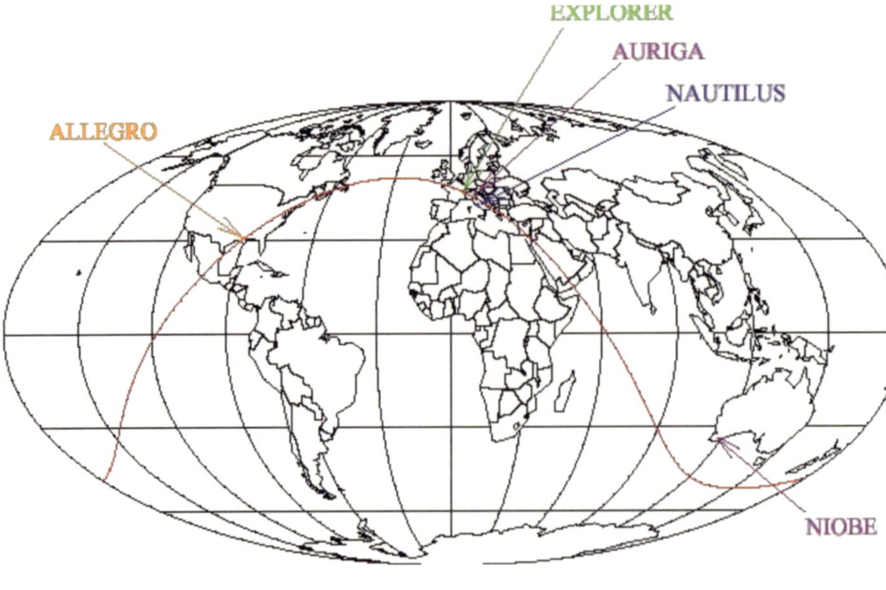

Figure 5.
Network of resonant-mass GW detectors that was operational for about two decades.

Monolithic sapphire parametric transducers can be used to improve the sensitivity of the resonant-mass GW detectors as can be seen in [18].

2.3 Pulsar timing

Pulsar timing is an indirect measurement of GW. Using signals that come from pulsars (a neutron star that rotates and sends, like a lighthouse, electromagnetic radiation at very regular intervals). Observatories around the world are trying to mine these signals to look for some small differences in these regular intervals, as can be seen in **Figure 6**. If a GW passes through this signal, the part of the signal orthogonal to the GW will arrive at Earth with a certain delay. The pulses are folded on top of each other, looking for some residuals. The problem is that they only work for big periods of time, like some decades. That is because these experiments detect signals in the range of nanoHertz. A review of these experiments can be seen in [19]. Some results are appearing in the literature, but with no concrete results yet.

There are three major experiments: Parkes Pulsar Timing Array (PPTA, Australia), European Pulsar Timing Array (EPTA, Europe) and NanoGrav (American project).

A review of these advanced ground based detectors can be seen in [20].

2.4 Polarization of cosmic microwave background

Another range to explore GW radiation is the ultra-low frequency domain around 10^{-18} Hz. Trying to find the effects of GW radiation imprinted in the cosmic

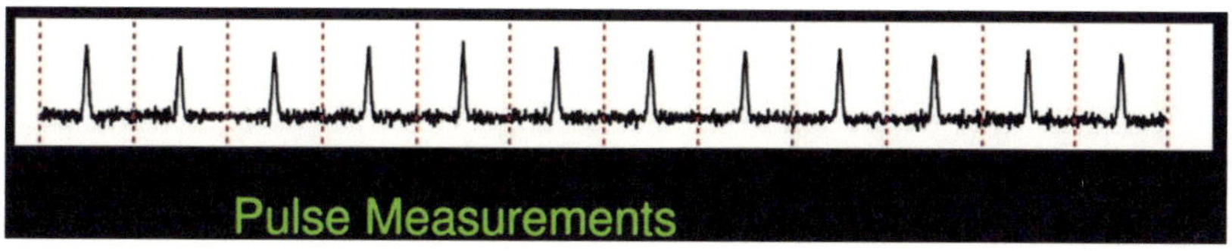

Figure 6.
Signal arriving at Earth coming from a pulsar.

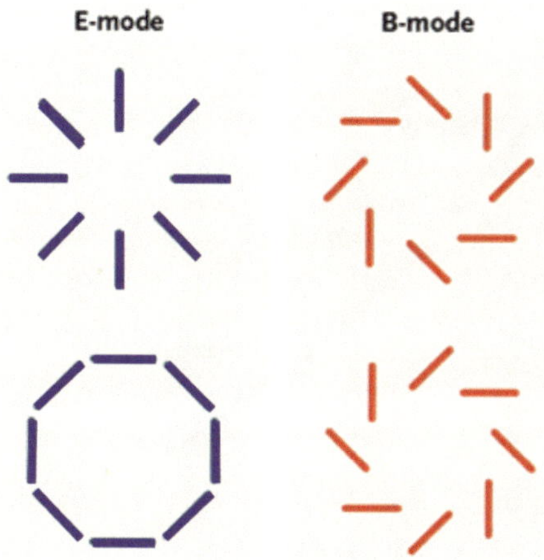

Figure 7.
Representation of electric and magnetic polarizations of the microwave background.

microwave background (CMB) when the universe was 300,000 years old. So far, no experiment has been able to measure the respective B-modes (magnetic modes) of the CMB, and more information can be obtained in [21]. Some researchers believe that because of the weakness of the signal, no signal could ever be detected. **Figure 7** shows the E and B-mode polarizations of the CMB.

2.5 Space interferometric detectors

Space interferometric detectors are interferometers mounted in satellites that are not affected by seismic noise, and, as they are much bigger, they are sensitive to frequencies much lower than their terrestrial relative. Some information about them and other future detectors can be found in [22]. As they are built in space, they are not limited by the seismic noise that limits the sensitivity of ground detectors.

A list of these experiments include: LISA (Laser Interferometer Space Antenna, NASA-ESA), DECIGO (Deci-Hertz Interferometric Gravitational Wave Observatory–Japan), TianQuin (China) and Taiji (China).

3. Experiments at all the frequency spectrum

Figure 8 summarizes all the expected sources and the expected main detectors for the observation of GW radiation in the audio regime (some kHz) and lower. The green area is the background of binary sources. Below it, with a lower amplitude, there is a region where is expected the presence of the GW relic background, fossil radiation from the very beginning of the universe, probably hidden because of the binary background.

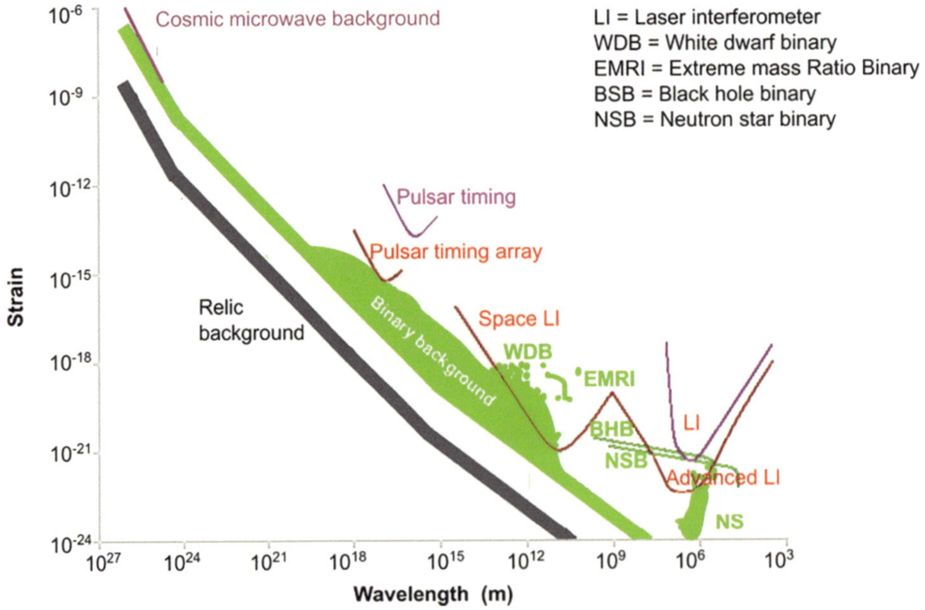

Figure 8.
Spectrum for the detection sources and detectors.

4. Conclusions

We are living in a very exciting moment in the history of science and astronomy, a new window for the observation of the universe is opening. With the new experiments coming in operation, all ranges of frequencies could be observed. This is for GW astronomy as the observations in visible, infrared, ultraviolet, x-rays and gamma-rays are for astronomy.

With all that information, some of the most hidden astrophysical phenomena could be finally explored. It will be possible to make gravitational observations together with electromagnetic observations. The first one, the kilonova observation with gravitational and gamma-rays gave a glimpse of the merge of two neutron stars, the gravitational signal gives the masses of the merging objects and the electromagnetic signal gives information about the electromagnetic emission of the process, a much more complete description.

With the new GW detector much more information can be gathered, with the space laser-interferometric detectors, the orbits of compact objects can be found, objects that do not emit light and are part of binary systems, and white dwarf binaries. With the pulsar timing, the merger of galaxies with their central black holes can be better understood and if the measurement of the primordial GW can be achieved, a glimpse of the very first instants of the universe can make what information it can bring.

These are really exciting times.

Acknowledgements

The author acknowledges Conselho de Pesquisa e Desenvolvimento Cientifico (CNPq, Brazil) for Grant number 312454/2021-0 and Fundacao de Amparo a Pesquisa do Estado de Sao Paulo (FAPESP) Grant number 2013/26258-4.

Conflict of interest

The author declares no conflict of interest.

Author details

Carlos Frajuca
Rio Grande Federal University, Rio Grande, RS, Brazil

*Address all correspondence to: frajuca@gmail.com

References

[1] Abbott BP et al. Observation of gravitational waves from a binary black hole merger. Physical Review Letters. 2016;**116**:061102. DOI: 10.1103/PhysRevLett.116.061102

[2] Abbott BP et al. GW170817: Observation of gravitational waves from a binary neutron star inspiral. Physical Review Letters. 2017;**119**:161101. DOI: 10.1103/PhysRevLett.119.161101

[3] Aguiar OD et al. The Schenberg spherical antenna: Status report. In: 12th Marcel Grossmann Meeting on Recent Developments in Theoretical and Experimental General Relativity, Astrophysics and Relativistic Field Theories - Proceedings of the MG 2009 Meeting on General Relativity. Singapore: World Scientific; 2012. pp. 1649-1651

[4] Cahillane C, Mansell G. Review of the advanced LIGO gravitational wave observatories leading to observing run four. ArXiv: 2202.00847v1 [gr-qc]. 2022. DOI: 10.48550/arXiv.2202.00847

[5] Nardecchia I. Detecting gravitational waves with advanced Virgo. Galaxies. 2022;**18**:28. DOI: 10.3390/galaxies10010028

[6] Somiya K. Detector configuration of KAGRA - The Japanese cryogenic gravitational-wave detector. ArXiv:1111.7185v2 [gr-qc]. 2011. DOI: 10.1088/0264-9381/29/12/124007

[7] Frajuca C, Selbach L, da Silva DA, de Sales DA, Ferrari F, Magalhaes NS. Coherent quantum states in resonant-mass gravitational wave detectors. Astronomische Nachrichten. 2023;**344**:1-2

[8] Prado ARC, Bortoli FS, Magalhaes NS, Duarte RN, Frajuca C, Souza RC. Modelling a mechanical antenna for a calibrator for interferometric gravitational wave detector using finite elements method. Journal of Physics: Conference Series. 2021;**2090**:012157

[9] Frajuca C, Prado ARC, Souza MA, Magalhaes NS. The challenge of calibrating a laser-interferometric gravitational wave detector. Astronomische Nachrichten. 2021;**342**(1-2):115-122

[10] Bortoli FS, Frajuca C, Magalhaes NS, de Souza ST, da Silva Junior WC. On the dilution refrigerator thermal connection for the SCHENBERG gravitational wave detector. Brazilian Journal of Physics. 2020;**50**(5):541-547

[11] da Silva Bortoli F, Frajuca C, Magalhaes NS, Aguiar OD, de Souza ST. On the cabling seismic isolation for the microwave transducers of the Schenberg detector. Brazilian Journal of Physics. 2019;**49**(1):133-139

[12] Hamilton WO. Performance of the ALLEGRO detector—And what our experience tells us about spherical detectors. In: Contribution to: OMNI 1: 1st International Workshop on an Omnidirectional Gravitational Radiation Observatory. Singapore: World Scientific; 1997. pp. 19-26

[13] da Silva Bortoli F, Frajuca C, de Sousa ST, de Waard A, Magalhaes NS, de Aguiar OD. On the massive antenna suspension system in the Brazilian gravitational wave detector Schenberg. Brazilian Journal of Physics. 2016;**46**(3):308-315

[14] Frajuca C, Ribeiro KL, Andrade LA, Aguiar OD, Magalhães NS, De Melo Marinho R Jr. A noise model

for the Brazilian gravitational wave detector 'Mario Schenberg'. Classical and Quantum Gravity. 2004;**21**(5):S1107-S1111

[15] Frajuca C, Ribeiro KL, Andrade LA, Velloso WF Jr, Melo JL, Aguiar OD, et al. Transducers for the Brazilian gravitational wave detector 'Mario Schenberg'. Classical and Quantum Gravity. 2002;**19**(7):1961-1965

[16] Magalhães NS, Johnson WW, Frajuca C, Aguiar OD. Determination of astrophysical parameters from the spherical gravitational wave detector data. Monthly Notices of the Royal Astronomical Society. 1995;**274**(3):670-678

[17] Frajuca C, Souza MA, Coppedé D, Nogueira PRM, Bortoli FS, Santos GA, et al. Optimization of a composite quadrupole mass at high-speed rotation. Journal of the Brazilian Society of Mechanical Sciences and Engineering. 2018;**40**(6):319

[18] Tobar M. Monolithic sapphire parametric transducer operation at cryogenic temperatures. The Review of Scientific Instruments. 2000;**71**:2737-2741

[19] Dahal PK. Review of pulsar timing array for gravitational wave research. arXiv.2002.01954v2 [astro-ph.IM]. 2020. DOI: 10.48550/arXiv.2002.01954

[20] Adier M, Aguilar F, Akutsu T, Arain MA, Ando M, Anghinolfi L, et al. Progress and challenges in advanced ground-based gravitational-wave detectors. General Relativity and Gravitation. 2014;**46**(8):1-22

[21] Kamionkowski M, Kovetz ED. The quest for B modes from inflationary gravitational waves. Annual Review of Astronomy and Astrophysics. 2016;**54**:227-269. DOI: 10.1146/annurev-astro-081915-023433

[22] Acernese F et al. Concepts and research for future detectors: Summary of the Amaldi 10 C4 session. General Relativity and Gravitation. 2014;**46**(5):1-20

Chapter 7

Effects of Gravitational Waves on Two-Level Atom Moving in a Quantized Traveling Light Field: Exact Solution via Path Integral

Hilal Benkhelil

Abstract

We adopt a coherent states path integral formalism to study the system of a two-level atom moving in a quantized traveling light field and a gravitational field. By using the phase space and some rotations in the space of coherent states, have enabled greatly simplify the calculations. The propagator is first written in a standard form, $\int \mathcal{D}(path) \exp(i/\hbar)S(path)$, by replacing the spin with a unit vector aligned along the polar and azimuthal directions. Then, it is determined exactly due to the auxiliary equation which has a spacial function as a solution. The corresponding wave functions have been deduced by applying the principles of quantum mechanics. The results obtained are perfectly identical with those found by other standard methods.

Keywords: path integral, propagator, coherent states, the Jaynes-Cummings model, gravitational waves

1. Introduction

Gravitational waves are waves of the intensity of gravity generated by the accelerated masses of an orbital binary system that propagate as waves outward from their source at the speed of light. Gravitational waves transport energy as gravitational radiation, a form of radiant energy similar to electromagnetic radiation [1].

The effect of gravitational waves on the movement of atoms is important and cannot be neglected. In atomic optics experiments have made it possible to create atomic clouds and beams with very small velocity [2]. For atoms moving with a velocity of a few millimeters or centimeters per second for a time period of several milliseconds or more, the influence of Earth's acceleration becomes important and cannot be neglected [3].

Among the simplest scheme to investigate the interaction between a two-level atom and a single-mode quantized electromagnetic field is the Jaynes-Cummings model (JCM) [4]. The JCM has received a great deal of experimental as well as theoretical attention [5–8]. Over the years, the JCM has been extended and generalized in many directions, for example, the effects of finite cavity damping [9, 10],

IntechOpen

intensity dependent coupling [11, 12] and the introduction of a Kerr-like medium [13]. A very significant and noteworthy generalization of JCM is to include the quantization of atomic momentum and position [14–16] so that the internal and external dynamics of the atom could be treated into this model.

For this reason, we are devoted to this model of interaction, we use the path integral formalism in the bosonic and fermionic coherent states representation to solve the generalized JCM in the presence of a gravitational field, governed by the Hamiltonian ($\hbar = 1$)

$$\begin{aligned} H = \frac{p^2}{2m} - mgr + \omega\left(a^\dagger a + \frac{1}{2}\right) \\ + \frac{\omega_a}{2}\boldsymbol{\sigma}_z + \lambda\left(e^{+ikr} a\boldsymbol{\sigma}_+ + e^{-ikr} a^\dagger \boldsymbol{\sigma}_-\right). \end{aligned} \tag{1}$$

Here, a and $a^\dagger$ are the atomic flipping operators, ω is the field frequency, λ is the atom-field coupling constant, ω_a is the transition frequency between the levels, $\boldsymbol{\sigma}_z, \boldsymbol{\sigma}_+, \boldsymbol{\sigma}_-$ are the usual Pauli matrices, k is the wave vector of the running wave, p and r denote, respectively, the momentum and position operators of the atomic center of mass motion, and g is Earth's gravitational acceleration.

In this paper, we propose to present an alternative solution to the given problem by the coherent states path integral representation via the Schwinger's model of spin. It should be noted that, in the semi-classical description of two-level atom interacting with electromagnetic wave, the effect of the gravitational field has been recently studied using fermionic coherent state path integral [17].

This paper is organized as follows. In Section 2, we give a brief review of the bosonic and fermionic coherent states path integral representation for our further computations. In Section 3, after setting up a path integral formalism for the propagator, we perform the direct calculations over the angular variables. Accordingly, the integration over the bosonic variables is easy to carry out and the result is given as a perturbation series, which is summed up exactly. The explicit result of the propagator is directly computed and the wave function is then deduced in Section 4. Finally, we present our conclusions in the last section, we give a brief review of the bosonic and fermionic coherent states path integral representation.

2. Coherent state propagator

At this stage, we briefly give the definitions and some properties related to bosonic and fermionic coherent states in the path integral formalism.

For the coherent states $|Z\rangle$ relative to bosons, the properties are known. These are

- the eigenstate of the annihilation operator $a([a, a^\dagger] = 1)$

$$a|Z\rangle = Z|Z\rangle, \tag{2}$$

- they can also be created from the vacuum state $|0\rangle$ by applying a unitary operator called the displacement operator

$$|Z\rangle = e^{Za^\dagger - Z^* a}|0\rangle, \tag{3}$$

In this case, the scalar product and the projector operator are respectively

$$\langle Z|Z'\rangle = \exp\left(Z^* Z' - \frac{1}{2}\left(|Z|^2 + |Z'|^2\right)\right), \tag{4}$$

$$\int \frac{d^2 Z}{\pi} |Z\rangle\ \langle Z| = 1. \tag{5}$$

As for spin interaction, we use the approach whose recipe includes replacing the Pauli matrices σ_i by a unit vector $\mathbf{n}$ directed along (θ, φ)

$$|\theta, \varphi\rangle = e^{-i\varphi S_z} e^{-i\theta S_y} |\uparrow\rangle \tag{6}$$

which is obtained by applying two rotations with angles θ and φ around the z and y axes on the state $|\uparrow\rangle$, and whose scalar product and projector are respectively

$$\langle \theta, \varphi | \theta', \varphi'\rangle = \cos\frac{\theta}{2}\cos\frac{\theta'}{2} e^{\frac{i}{2}(\varphi - \varphi')} + \sin\frac{\theta}{2}\sin\frac{\theta'}{2} e^{-\frac{i}{2}(\varphi-\varphi')} \tag{7}$$

$$\frac{1}{2\pi}\int d\varphi d\cos(\theta) |\theta, \varphi\rangle\langle\theta, \varphi| = \mathbf{I}. \tag{8}$$

Taking into account that

$$\langle \theta, \varphi | \sigma_z | \theta_n, \varphi'\rangle = \cos\frac{\theta}{2}\cos\frac{\theta'}{2} e^{+\frac{i}{2}(\varphi - \varphi')} - \sin\frac{\theta}{2}\sin\frac{\theta'}{2} e^{-\frac{i}{2}(\varphi-\varphi')}, \tag{9}$$

$$\langle \theta, \varphi | \sigma_+ | \theta_n, \varphi'\rangle = \cos\frac{\theta}{2}\sin\frac{\theta'}{2} e^{+\frac{i}{2}(\varphi + \varphi')}, \tag{10}$$

$$\langle \theta, \varphi | \sigma_- | \theta, \varphi'\rangle = \sin\frac{\theta}{2}\cos\frac{\theta'}{2} e^{-\frac{i}{2}(\varphi + \varphi')}. \tag{11}$$

According to the habitual construction procedure of the path integral. We consider the quantum state $|r, Z, \theta, \varphi\rangle$, where Z is a complex variable generating the dynamics of the field, (θ, φ) are the polar angles variables generating the dynamics of the spin and r the real variable describing the atom position, with the corresponding projector

$$\int |r\rangle\langle r| dr^3 = 1. \tag{12}$$

The transition amplitude from the initial state $|r_i, Z_i, \theta_i, \varphi_i\rangle$ and the final state $\left|r_f, Z_f, \theta_f, \varphi_f\right\rangle$ at $t_i = 0$ to the final state at $t_f = T$ is defined with the matrix elements of the evolution operator

$$K(f, i; T) = \left\langle r_f, Z_f, \theta_f, \varphi_f | U(T) | r_i, Z_i, \theta_i, \varphi_i \right\rangle, \tag{13}$$

where

$$U(T) = \mathbf{T}_D \exp\left(-\frac{i}{\hbar}\int_0^T H(t)dt\right), \tag{14}$$

with T_D the Dyson chronological operator.

For moving to the path-integral representation, we first subdivide the time interval $[0, T]$ into $N + 1$ intermediate moments of length ε. Using the Trotter's formula, we then introduce in (13) the projectors according to these N intermediate instants, which are regularly distributed between 0 and T. We obtain the discrete path-integral form of the propagator

$$
\begin{aligned}
K(f,i;T) = \lim_{N\to\infty}\left(\frac{m}{2\pi i\varepsilon}\right)^{3N/2} &\int \prod_{n=1}^{N} d^3 r_n \int \prod_{n=1}^{N} \frac{d^2 Z_n}{\pi} e^{-Z_n^* Z_n} \\
&\times \prod_{n=1}^{N+1} \langle r_n, Z_n | e^{-i\varepsilon H_0} | r_{n-1}, Z_{n-1} \rangle \qquad (15)\\
\times \lim_{N\to\infty} \int \prod_{n=1}^{N} \frac{d\varphi d\cos(\theta)}{2\pi} &\prod_{n=1}^{N+1} \langle Z_n, \theta_n, \varphi_n | e^{-i\varepsilon H_{int}} | Z_{n-1}, \theta_{n-1}, \varphi_{n-1} \rangle,
\end{aligned}
$$

where

$$
\begin{cases}
r_{N+1} = r_f, Z_{N+1} = Z_f, \theta_{N+1} = \theta_f, \varphi_{N+1} = \varphi_f \\
\text{and} \\
r_0 = r_i, Z_0 = Z_i, \theta_0 = \theta_i, \varphi_0 = \varphi_i
\end{cases}, \qquad (16)
$$

and the propagator related to our problem takes the form of Feynman path integral

$$
K = \int \mathcal{D}_{path} e^{i(Action)}. \qquad (17)
$$

After having obtained the conventional form, it remains to integrate it, in order to extract the interesting physical properties. We thus proceed to the calculation of $K(f, i; T)$ in the next section.

3. The propagator calculation

We note that (15) is written like the following discrete-time form

$$
\begin{aligned}
K(f,i;T) = \lim_{N\to\infty}\left(\frac{m}{2\pi i\varepsilon}\right)^{3N/2} &\int \prod_{n=1}^{N} d^3 r_n \prod_{n=1}^{N+1} \exp\left\{i\left[\frac{m}{2\varepsilon}(r_n - r_{n-1})^2 + \varepsilon m g r_n\right]\right\} \\
\times \lim_{N\to\infty} \frac{1}{2\pi i} &\int \prod_{n=1}^{N} dZ_n^* dZ_n e^{-Z_n^* Z_n} \prod_{n=1}^{N+1} \exp\left\{\frac{i\varepsilon}{\hbar}\omega\left(Z_n^* Z_{n-1} + \frac{1}{2}\right)\right\} \\
&\times \lim_{N\to\infty} \int \prod_{n=1}^{N} \frac{d\varphi d\cos(\theta)}{2\pi} \prod_{n=1}^{N+1} \\
\times \left(\cos\frac{\theta_n}{2} e^{+\frac{i}{2}\varphi_n}, \cos\frac{\theta_n}{2} e^{-\frac{i}{2}\varphi_n}\right) &R(r_n, t_n) \left(\cos\frac{\theta_{n-1}}{2} e^{-\frac{i}{2}\varphi_{n-1}} \sin\frac{\theta_{n-1}}{2} e^{+\frac{i}{2}\varphi_{n-1}}\right),
\end{aligned}
\qquad (18)
$$

with

$$R(r_n, t_n) = \left[e^{-i\varepsilon\frac{\omega_a}{2}\sigma_z} + i\varepsilon K(Z_n, r_n, t_n)\right], \tag{19}$$

where

$$K(r_n, t_n) = \begin{pmatrix} 0 & -\lambda e^{+ikr_n} Z_n \\ -\lambda e^{-ikr_{n-1}} Z_n^* & 0 \end{pmatrix}. \tag{20}$$

To integrate, it is necessary to eliminate first the inconvenient terms $e^{\pm ikr}$, which appear in the action with the help of the following change

$$\varphi_n = \varphi'_n + kr_n. \tag{21}$$

The new expression we have to calculate is therefore

$$\begin{aligned} K(f, i; T) = {} & \lim_{N\to\infty}\left(\frac{m}{2\pi i\varepsilon}\right)^{3N/2} \int \prod_{n=1}^{N} d^3 r_n \prod_{n=1}^{N+1} \exp\left\{ i\left[\frac{m}{2\varepsilon}(r_n - r_{n-1})^2 + \varepsilon m g r_n\right]\right\} \\ & \times \lim_{N\to\infty} \frac{1}{2\pi i} \int \prod_{n=1}^{N} dZ_n^* dZ_n e^{-Z_n^* Z_n} \prod_{n=1}^{N+1} \exp\left\{\frac{i\varepsilon}{\hbar}\omega\left(Z_n^* Z_{n-1} + \frac{1}{2}\right)\right\} \\ & \times \lim_{N\to\infty} \int \prod_{n=1}^{N} \frac{d\varphi' d\cos(\theta)}{2\pi} \prod_{n=1}^{N+1} \\ & \times \left(\cos\frac{\theta_n}{2} e^{+\frac{i}{2}\varphi'_n}, \cos\frac{\theta_n}{2} e^{-\frac{i}{2}\varphi'_n}\right) R_1(r_n, t_n) \begin{Bmatrix} \cos\frac{\theta_{n-1}}{2} e^{-\frac{i}{2}\varphi'_{n-1}} \\ \sin\frac{\theta_{n-1}}{2} e^{+\frac{i}{2}\varphi'_{n-1}} \end{Bmatrix}, \end{aligned} \tag{22}$$

where

$$R_1(r_n, t_n) = \begin{pmatrix} \left(1 - i\varepsilon\frac{\omega_{eg}}{2}\right) e^{ik\Delta r_n} & -i\varepsilon\lambda Z_n \\ -i\varepsilon\lambda Z_n^* & 1 + i\varepsilon\frac{\omega_{eg}}{2} \end{pmatrix} \tag{23}$$

$$\text{with, } \Delta r_n = r_n - r_{n-1}. \tag{24}$$

We use the following identity

$$\int_{-\infty}^{+\infty} \frac{d^3 p_n}{(2\pi)^3} \exp\left[\frac{-i\varepsilon}{2m} p_n^2 + i p_n \Delta r_n\right] = \left(\frac{m}{2\pi i\varepsilon}\right)^{3/2} \exp\frac{im}{2\varepsilon}(\Delta r_n)^2, \tag{25}$$

Thus, propagator (22) can be rewritten as

$$
\begin{aligned}
K(f,i;T) = {} & \lim_{N\to\infty} \int \prod_{n=1}^{N} d^3 r_n \int_{-\infty}^{+\infty} \frac{d^3 p_n}{(2\pi)^3} \exp \sum_{n=1}^{N+1} \left[\frac{-i\varepsilon}{2m} p_n^2 + i p_n \Delta r_n + i\varepsilon m g r_n \right] \\
& \times \lim_{N\to\infty} \frac{1}{2\pi i} \int \prod_{n=1}^{N} dZ_n^* \, dZ_n e^{-Z_n^* Z_n} \prod_{n=1}^{N} \exp\left\{ -i\varepsilon\omega \left(Z_n^* Z_{n-1} + \frac{1}{2} \right) \right\} \\
& \times \lim_{N\to\infty} \int \prod_{n=1}^{N} \frac{d\varphi' d\cos(\theta)}{2\pi} \prod_{n=1}^{N+1} \\
& \times \left(\cos\frac{\theta_n}{2} e^{+\frac{i}{2}\varphi'_n}, \cos\frac{\theta_n}{2} e^{-\frac{i}{2}\varphi'_n} \right) R_2(p_n, t_n) \begin{Bmatrix} \cos\frac{\theta_{n-1}}{2} e^{-\frac{i}{2}\varphi'_{n-1}} \\ \sin\frac{\theta_{n-1}}{2} e^{+\frac{i}{2}\varphi'_{n-1}} \end{Bmatrix},
\end{aligned}
\tag{26}
$$

with

$$
R_2(p_n, t_n) = \begin{pmatrix} 1 - i\varepsilon\left(\frac{\omega_a}{2} + \frac{k^2}{2m} + \frac{k p_n}{m} \right) & -i\varepsilon\lambda Z_n \\ -i\varepsilon\lambda Z_n^* & 1 + i\varepsilon\frac{\omega_a}{2} \end{pmatrix}. \tag{27}
$$

By integrating over the N variables r_n, we clearly get Dirac functions $\delta(\dot{p} - mg)$, which means that the particle is only subject to the action of gravitational waves. The atom impulsion is

$$
p_n = m g t_n + p_0 \quad \text{where} \quad (p_0 \text{ constant}). \tag{28}
$$

The contribution of the time-linear function in the computation of the propagator has the following result

$$
\begin{aligned}
K(f,i;T) = {} & \int \frac{d^3 p_0}{(2\pi)^3} e^{i(mgt + p_0) r|_0^T} e^{-i\left(\frac{m}{6} g^2 T^3 + \frac{p_0^2}{2m} T + \frac{1}{2} g p_0 T^2 \right)} \\
& \times \int \frac{dZ_n^* dZ_n}{2\pi i} e^{-Z_n^* Z_n} \exp\left\{ i \int_0^T dt \left[\frac{i}{2} \left(\dot{Z}^* Z - \dot{Z} Z^* \right) - \omega Z^* Z - \frac{\omega}{2} \right] \right\} \\
& \times \lim_{N\to\infty} \int \prod_{n=1}^{N} \frac{d\varphi' d\cos(\theta)}{2\pi} \prod_{n=1}^{N+1} \\
& \times \left(\cos\frac{\theta_n}{2} e^{+\frac{i}{2}\varphi'_n}, \cos\frac{\theta_n}{2} e^{-\frac{i}{2}\varphi'_n} \right) R_3(p_0, t_n) \begin{Bmatrix} \cos\frac{\theta_{n-1}}{2} e^{-\frac{i}{2}\varphi'_{n-1}} \\ \sin\frac{\theta_{n-1}}{2} e^{+\frac{i}{2}\varphi'_{n-1}} \end{Bmatrix},
\end{aligned}
\tag{29}
$$

with

$$
R_3(p_0, t_n) = \begin{pmatrix} 1 - i\varepsilon\left(\frac{\omega_a}{2} + \frac{k^2}{2m} + k g t_n + \frac{k p_0}{m} \right) & -i\varepsilon\lambda Z_n \\ -i\varepsilon\lambda Z_n^* & 1 + i\varepsilon\frac{\omega_a}{2} \end{pmatrix}. \tag{30}
$$

At this level, let us deal with the integration over the Grassmann variables. We shall introduce the following change

$$\begin{Bmatrix} \cos\dfrac{\theta_{n-1}}{2}e^{-\frac{i}{2}\varphi'_{n-1}} \\ \sin\dfrac{\theta_{n-1}}{2}e^{+\frac{i}{2}\varphi'_{n-1}} \end{Bmatrix} = e^{-\frac{i}{2}\left(\frac{k^2}{2m}t_n+\frac{1}{2}kgt_n^2+\frac{kp_0}{m}t_n\right)} \begin{Bmatrix} \cos\dfrac{\theta'_{n-1}}{2}e^{-\frac{i}{2}\varphi''_{n-1}} \\ \sin\dfrac{\theta'_{n-1}}{2}e^{+\frac{i}{2}\varphi''_{n-1}} \end{Bmatrix}, \tag{31}$$

so the expression of the propagator become

$$\begin{aligned} K(f,i;T) = &\int \frac{d^3p_0}{(2\pi)^3} e^{i(mgt+p_0)r|_0^T} e^{-i\left(\frac{m}{6}g^2T^3+\frac{p_0^2}{2m}T+\frac{1}{2}gp_0T^2\right)} \\ &\times \int \frac{dZ_n^* dZ_n}{2\pi i} e^{-Z_n^* Z_n} \exp\left\{ i\int_0^T dt \left[\frac{i}{2}\left(\dot{Z}^* Z - \dot{Z}Z^*\right) - \omega Z^* Z - \frac{\omega}{2} \right] \right\} \\ &\times \lim_{N\to\infty} \int \prod_{n=1}^{N} \frac{d\varphi'' d\cos(\theta')}{2\pi} \prod_{n=1}^{N+1} \\ &\times \left(\cos\frac{\theta'_n}{2}e^{+\frac{i}{2}\varphi''_n}, \cos\frac{\theta'_n}{2}e^{-\frac{i}{2}\varphi''_n} \right) R_4(p_0,\Omega_n) \begin{Bmatrix} \cos\dfrac{\theta'_{n-1}}{2}e^{-\frac{i}{2}\varphi''_{n-1}} \\ \sin\dfrac{\theta'_{n-1}}{2}e^{+\frac{i}{2}\varphi''_{n-1}} \end{Bmatrix}, \end{aligned} \tag{32}$$

where

$$R_4(p_0,\Omega_n) = \begin{pmatrix} 1-\dfrac{i}{2}\varepsilon\Omega(p,g,t) & -i\varepsilon\lambda Z_n \\ -i\varepsilon\lambda Z_n^* & 1+\dfrac{i}{2}\varepsilon\Omega(p,g,t) \end{pmatrix}, \tag{33}$$

with

$$\Omega(p,g,t) = \omega_a + \frac{k^2}{2m} + kgt_n + \frac{kp_0}{m}. \tag{34}$$

We integrate over all variables Z_n, (32) become

$$\begin{aligned} K(f,i;T) = &\int \frac{d^3p_0}{(2\pi)^3} e^{i(mgt+p_0)r|_0^T} e^{-i\left(\frac{m}{6}g^2T^3+\frac{p_0^2}{2m}T+\frac{1}{2}gp_0T^2\right)} \\ &\times \lim_{N\to\infty} \int \prod_{n=1}^{N} \frac{d\varphi'' d\cos(\theta')}{2\pi} \prod_{n=1}^{N+1} \\ &\times \left(\cos\frac{\theta'_n}{2}e^{+\frac{i}{2}\varphi''_n}, \cos\frac{\theta'_n}{2}e^{-\frac{i}{2}\varphi''_n} \right) K(p_0,\Omega_n) \begin{Bmatrix} \cos\dfrac{\theta'_{n-1}}{2}e^{-\frac{i}{2}\varphi''_{n-1}} \\ \sin\dfrac{\theta'_{n-1}}{2}e^{+\frac{i}{2}\varphi''_{n-1}} \end{Bmatrix}, \end{aligned} \tag{35}$$

where

$$K(p_0,\Omega_n) = e^{-\frac{|z_f|^2+|z_i|^2}{2}} \sum_{j=0}^{\infty} \left\{ \frac{Z_f^{*j} Z_i^j}{j!} e^{-i\left(j\omega_c T + \frac{1}{2}\omega_c T + \frac{1}{2}\int_0^T dt\Omega(p,g,t)\right)} \right\} \times \begin{pmatrix} 1 - i\varepsilon\Delta(p,g,t) & -i\varepsilon\lambda\sqrt{j+1} \\ -i\varepsilon\lambda\sqrt{j+1} & 1 + i\varepsilon\Delta(p,g,t) \end{pmatrix}, \quad (36)$$

where

$$\Delta(t_j) = \frac{1}{2}\left[\omega - \left(\omega_a + 3\frac{k^2}{2m} + kgt_j + \frac{kp_0}{m}\right)\right], \quad (37)$$

is the detuning of the atom-field interaction which depends on both the atomic momentum and the gravitational field.

Now, to calculate propagator (35) let us introduce the complex variable z [18].

$$z = \tan\frac{\theta'}{2} e^{+i\varphi''} \text{ and } |\theta', \varphi''\rangle = \left(\cos\frac{\theta'_{n-1}}{2} e^{-\frac{i}{2}\varphi''_{n-1}} \sin\frac{\theta'_{n-1}}{2} e^{+\frac{i}{2}\varphi''_{n-1}}\right) = e^{-\frac{i}{2}\varphi''_{n-1}} |z\rangle \quad (38)$$

The propagator in the z representation is

$$K(f,i;T) = \int \frac{d^3 p_0}{(2\pi)^3} e^{i(mgt+p_0)r|_0^T} e^{-i\left(\frac{m}{6}g^2T^3 + \frac{p_0^2}{2m}T + \frac{1}{2}gp_0T^2\right)} \times e^{-\frac{|z_f|^2+|z_i|^2}{2}} \sum_{j=0}^{\infty} \left\{ \frac{Z_f^{*j} Z_i^j}{j!} \exp\left[-i\left(j\omega_c T + \frac{1}{2}\omega_c T + \frac{1}{2}\int_0^T dt\Omega(p,g,t)\right)\right]\right\} \times e^{+\frac{i}{2}\varphi''_f} e^{-\frac{i}{2}\varphi''_i} \langle z_f| \exp\left[-i\int_{\Delta_0}^{\Delta(T)} H(p_0,\Delta) dS\right] |z_i\rangle, \quad (39)$$

where $H(p_0,\Delta)$ belongs to $SU(2)$ algebra

$$H(p_0,\Delta) = \Delta_n S_0 + f_n S_+ + f_n S_- \quad (40)$$

and

$$f = \lambda\sqrt{j+1}. \quad (41)$$

The computation of propagator (39) is readily and the results are given by

$$K(f,i;T)=\int\frac{d^3p_0}{(2\pi)^3}e^{i(mgt+p_0)r|_0^T}e^{-i\left(\frac{m}{6}g^2T^3+\frac{p_0^2}{2m}T+\frac{1}{2}gp_0T^2\right)}$$
$$\times e^{-\frac{|z_f|^2+|z_i|^2}{2}}\sum_{j=0}^{\infty}\left\{\frac{Z_f^{*j}Z_i^j}{j!}\exp\left[-i\left(j\omega_cT+\frac{1}{2}\omega_cT+\frac{1}{2}\int_0^T dt\Omega(p,g,t)\right)\right]\right\} \tag{42}$$
$$\times e^{+\frac{i}{2}\varphi_f''}e^{-\frac{i}{2}\varphi_i''}\langle z_f|\exp\left[\frac{a^*(\Delta)-b^*(\Delta)z_f+b(\Delta)z_i^*+a^*(\Delta)z_i^*z_f}{\left(1+|z_i|^2\right)^{\frac{1}{2}}\left(1+|z_f|^2\right)^{\frac{1}{2}}}\right]|z_i\rangle,$$

where $a(\Delta)$, and $b(\Delta)$ satisfy

$$\frac{da}{d\Delta}=-ifbe^{i\Delta}\quad,\quad\frac{db}{d\Delta}=-ifae^{-i\Delta} \tag{43}$$

with the boundary conditions

$$a(\Delta_0)=1\ ,\ \ b(\Delta_0)=0\quad,\quad\Delta_0=\frac{1}{2}\left[\omega-\left(\omega_a+3\frac{k^2}{2m}+\frac{kp_0}{m}\right)\right]. \tag{44}$$

In terms of the angular variables it becomes

$$K(f,i;T)=\int\frac{d^3p_0}{(2\pi)^3}e^{i(mgt+p_0)r|_0^T}e^{-i\left(\frac{m}{6}g^2T^3+\frac{p_0^2}{2m}T+\frac{1}{2}gp_0T^2\right)}$$
$$\times e^{-\frac{|z_f|^2+|z_i|^2}{2}}\sum_{j=0}^{\infty}\left\{\frac{Z_f^{*j}Z_i^j}{j!}\exp\left[-i\left(j\omega_cT+\frac{1}{2}\omega_cT+\frac{1}{2}\int_0^T dt\Omega(p,g,t)\right)\right]\right\} \tag{45}$$
$$\times\left(\cos\frac{\theta_f'}{2}e^{+\frac{i}{2}\varphi_f''},\ \cos\frac{\theta_f'}{2}e^{-\frac{i}{2}\varphi_f''}\right)\begin{pmatrix}a(T) & b(T)\\ -b^*(T) & a^*(T)\end{pmatrix}\left\{\begin{matrix}\cos\frac{\theta_i'}{2}e^{-\frac{i}{2}\varphi_i''}\\ \sin\frac{\theta_i'}{2}e^{+\frac{i}{2}\varphi_i''}\end{matrix}\right\},$$

where

$$a(T)=e^{i\Delta T}\left[C(1)H_{A_j}(B(T))+C(2)_1F_1\left(-\frac{A_j}{2};\frac{1}{2};B^2(T)\right)\right], \tag{46}$$

$$b(T)=\left[C(1)H_{A_j+1}(B(T))+C(2)_1F_1\left(-\frac{A_j+1}{2};\frac{1}{2};B^2(T)\right)\right], \tag{47}$$

where $H_n(x)$ and ${}_1F_1(\alpha;\beta;\gamma)$ are the Hermite and the confluent hypergeometric functions. Furthermore, we have

$$B(t)=(i+1)\left(\frac{\sqrt{2}}{2}kgt-\frac{\sqrt{2}}{4\sqrt{kg}}\Delta_0\right),\text{and }A_j=-(2+i\beta), \tag{48}$$

$$\text{with, }\beta=\left(\frac{y_j(p,g)-\Delta_0^2}{2kg}\right),\text{and, }y_j(p,g)=\sqrt{y_j(p,0)^2+2ikg}, \tag{49}$$

with $y_j(p,0)^2 = f^2 + \Delta_0^2$ as the gravity-dependent Rabi frequency. And

$$C(1) = \frac{c_1}{c}\ , \ C(2) = \frac{c_2}{c}, \tag{50}$$

with

$$c_1 = {}_1F_1\left(-\frac{1}{2}(A_j+1); \frac{1}{2}; B^2(0)\right), \tag{51}$$

and

$$c_2 = H_{(A_j+1)}(B(0)), \tag{52}$$

and

$$c = H_{A_j}(B(0))\,{}_1F_1\left(-\frac{1}{2}(A_j+1); \frac{1}{2}; B^2(0)\right) - H_{(A_j+1)}(B(0))\,{}_1F_1\left(-\frac{A_j}{2}; \frac{1}{2}; B^2(0)\right), \tag{53}$$

Now we come back to the old Grassmann variables (θ, φ). So, the exact expression of the propagator concerning to our problem is the following

$$\begin{aligned} K(f,i;T) = & \int \frac{d^3p_0}{(2\pi)^3} e^{i(mgt+p_0)r|_0^T} e^{-i\left(\frac{m}{6}g^2T^3 + \frac{p_0^2}{2m}T + \frac{1}{2}gp_0T^2\right)} \\ & \times e^{-\frac{|Z_f|^2+|Z_i|^2}{2}} \sum_{j=0}^{\infty} \left\{ \frac{Z_f^{*j} Z_i^j}{j!} e^{-i\left(j\omega_c T + \frac{1}{2}\omega_c T + \frac{1}{2}\int_0^T dt\Omega(p,g,t)\right)} \right\} \\ & \times \left(\cos\frac{\theta_f}{2} e^{+\frac{i}{2}\varphi_f}, \cos\frac{\theta_f}{2} e^{-\frac{i}{2}\varphi_f}\right) S(T) \begin{Bmatrix} \cos\frac{\theta_i}{2} e^{-\frac{i}{2}\varphi_i} \\ \sin\frac{\theta_i}{2} e^{+\frac{i}{2}\varphi_i} \end{Bmatrix} \end{aligned} \tag{54}$$

where

$$\begin{aligned} S(T) = & e^{-\frac{i}{2}\left(\frac{k^2}{2m}T + \frac{1}{2}kgT^2 + \frac{kp_0}{m}T\right)} \\ & \times \begin{pmatrix} e^{ikr_f} & 0 \\ 0 & 1 \end{pmatrix} \begin{pmatrix} a(T) & b(T) \\ -b^*(T) & a^*(T) \end{pmatrix} \begin{pmatrix} e^{-ikr_i} & 0 \\ 0 & 1 \end{pmatrix}. \end{aligned} \tag{55}$$

Noting that the angles θ, φ are allowed to vary in the limited domains $[0, 2\pi]$ and $[0, 4\pi]$, our final result for the propagator is thus the following:

$$\begin{aligned} K(f,i;T) &= \sum_{n=-\infty}^{+\infty} K\left(Z_f, \theta_f + 2n\pi, \varphi_f + 4n\pi, Z_i, \theta_i, \varphi_i; T\right) \\ &= K\left(Z_f, \theta_f, \varphi_f; Z_i, \theta_i, \varphi_i; T\right). \end{aligned} \tag{56}$$

Our problem is thus solved. We can then determine the corresponding wave functions.

4. Wave functions

Let us now eliminate the coherent states by computing the transition amplitudes between the proper states of the spin. We take as an example the matrix element

$$K_{\uparrow\uparrow}(Z_f, Z_i; T) = \langle\uparrow|K(Z_f, Z_i; T)|\uparrow\rangle. \tag{57}$$

With the help of the completeness relations, this amplitude becomes

$$\int \frac{d\cos(\theta_f)d\varphi_f}{2\pi}\left|\theta_f, \varphi_f\right\rangle\left\langle\theta_f, \varphi_f\right| = 1, \qquad \int \frac{d\cos(\theta i)d\varphi_i}{2\pi}|\theta_i, \varphi_i\rangle\langle\theta_i, \varphi_i| = 1 \tag{58}$$

Then

$$\begin{aligned} K_{\uparrow\uparrow}(Z_f, Z_i; T) &= \int \frac{d^3p_0}{(2\pi)^3} e^{i(mgt+p_0)r|_0^T} e^{-i\left(\frac{m}{6}g^2T^3+\frac{p_0^2}{2m}T+\frac{1}{2}gp_0T^2\right)} \\ &\times e^{-\frac{|z_f|^2+|z_i|^2}{2}} \sum_{j=0}^{\infty}\left\{\frac{Z_f^{*j}Z_i^j}{j!}\exp\left[-i\left(j\omega_cT+\frac{1}{2}\omega_cT+\frac{1}{2}\int_0^T dt\Omega(p,g,t)\right)\right]\right\} \\ &\times \int \frac{d\cos(\theta_f)d\varphi_f}{2\pi}\frac{d\cos(\theta_i)d\varphi_i}{2\pi}\left\langle\uparrow|\theta_f, \varphi_f\right\rangle \begin{pmatrix} S_{11}(T) & S_{12}(T) \\ S_{21}(T) & S_{22}(T) \end{pmatrix} \langle\theta_i, \varphi_i|\uparrow\rangle \end{aligned} \tag{59}$$

where

$$\left\langle\uparrow|\theta_f, \varphi_f\right\rangle = \cos\frac{\theta_f}{2}e^{-\frac{i}{2}\varphi_f}, \text{ and } \langle\theta_i, \varphi_i|\uparrow\rangle = \cos\frac{\theta_i}{2}e^{+\frac{i}{2}\varphi_i} \tag{60}$$

The integration over polar angles, the propagator matrix element for the up-up states is finally written

$$K_{\uparrow\uparrow}(Z_f, Z_i; T) = \int \frac{d^3p_0}{(2\pi)^3} e^{i(mgt+p_0)r|_0^T} e^{-i\left(\frac{m}{6}g^2T^3+\frac{p_0^2}{2m}T+\frac{1}{2}gp_0T^2\right)} S_{11}(T). \tag{61}$$

Repeating the calculations and considering all initial and final states of the spin, the propagator takes the following matrix form

$$\begin{aligned} K(m_f, m_i; T) &= \int \frac{d^3p_0}{(2\pi)^3} e^{i(mgt+p_0)r|_0^T} e^{-i\left(\frac{m}{6}g^2T^3+\frac{p_0^2}{2m}T+\frac{1}{2}gp_0T^2\right)} \\ &\times e^{-\frac{|z_f|^2+|z_i|^2}{2}} \sum_{j=0}^{\infty}\left\{\frac{Z_f^{*j}Z_i^j}{j!} e^{-i\left(j\omega_cT+\frac{1}{2}\omega_cT+\frac{1}{2}\int_0^T dt\Omega(p,g,t)\right)}\right\} \\ &\times \begin{pmatrix} S_{11}(T) & S_{12}(T) \\ S_{21}(T) & S_{22}(T) \end{pmatrix}. \end{aligned} \tag{62}$$

In order to extract the wave functions, it is more convenient to use the basis $|n\rangle$, where n is the occupancy number related to $|Z\rangle$ through

$$|Z\rangle = \exp\left(-\frac{|Z|^2}{2}\right)\sum_{n=0}^{\infty}\frac{Z^n}{\sqrt{n!}}|n\rangle. \tag{63}$$

Then the evolution operator is

$$\hat{U}(p,g;T) = e^{-iHt} = \begin{pmatrix} \Lambda_{\uparrow\uparrow} & \Lambda_{\uparrow\downarrow} \\ \Lambda_{\downarrow\uparrow} & \Lambda_{\downarrow\downarrow} \end{pmatrix}, \tag{64}$$

which is related to $K(f,i;T)$ through

$$e^{-iHt} = \int\frac{d^2Z_f}{\pi}\int\frac{d^2Z_i}{\pi}|Z_f\rangle\ K(Z_f,Z_i;T)\langle Z_i|. \tag{65}$$

Performing the integrations yields the matrix elements of the evolution operator [19].

$$\begin{aligned}\Lambda_{\uparrow\uparrow} =& f(p,g,t)e^{ik(r_f-r_i)}\\ \times\sum_{j=0}^{\infty}&e^{i\Delta T}\left[C(1)H_{A_j}(B(T)) + C(2)_1F_1\left(-\frac{A_j}{2};\frac{1}{2};B^2(T)\right)\right]|j\rangle\langle j|,\end{aligned} \tag{66}$$

$$\begin{aligned}\Lambda_{\downarrow\uparrow} =& f(p,g,t)e^{-ikr_i}\\ \times\sum_{j=0}^{\infty}&\left[C(1)H_{A_j+1}(B(T)) + C(2)_1F_1\left(-\frac{A_j+1}{2};\frac{1}{2};B^2(T)\right)\right]|j+1\rangle\langle j|,\end{aligned} \tag{67}$$

$$\begin{aligned}\Lambda_{\uparrow\downarrow} =& f(p,g,t)e^{ikr_f}\\ \times\sum_{j=0}^{\infty}&\left[C(1)H_{A_j+1}(B(T)) + C(2)_1F_1\left(-\frac{A_j+1}{2};\frac{1}{2};B^2(T)\right)\right]^*|j\rangle\langle j+1|,\end{aligned} \tag{68}$$

$$\begin{aligned}\Lambda_{\downarrow\downarrow} =& f(p,g,t)\\ \times\sum_{j=0}^{\infty}&e^{-i\Delta T}\left[C(1)H_{A_j}(B(T)) + C(2)_1F_1\left(-\frac{A_j}{2};\frac{1}{2};B^2(T)\right)\right]^*|j+1\rangle\langle j+1|,\end{aligned} \tag{69}$$

where

$$\begin{aligned}f(p,g,t) = &\int\frac{d^3p_0}{(2\pi)^3}e^{i(mgt+p_0)r}\Big|_0^T e^{-i\left(\frac{m}{6}g^2T^3+\frac{p_0^2}{2m}T+\frac{1}{2}gp_0T^2\right)}\\ &\times e^{-i\left[j\omega_cT+\frac{1}{2}\omega_cT+\frac{1}{2}\left(\omega_{eg}T+\frac{k^2}{2m}T+\frac{1}{2}kgT^2+\frac{kp_0}{m}T\right)\right]}\\ &\times e^{-\frac{i}{2}\left(\frac{k^2}{2m}T+\frac{1}{2}kgT^2+\frac{kp_0}{m}T\right)}.\end{aligned} \tag{70}$$

For the considered atomic system, the wave function is written in the following form

$$\Psi(p,g;T) = \int d^3p\hat{U}(p,g;T)\Psi(p,g,0), \tag{71}$$

where

$$\Psi(p,g;0) = \int d^3p\varphi(p)\sum_{n=0}\omega_n\begin{pmatrix}|n\rangle\\0\end{pmatrix}, \tag{72}$$

with

$$\omega_n = \frac{Z^n}{\sqrt{n!}}e^{-\frac{|Z|^2}{2}}, \tag{73}$$

can be deduced, they become equal to

$$\Psi(p,g;T) = \sum_{j=0}\sum_{n=0}\omega_n f(p,g,t)$$
$$\times\begin{pmatrix} e^{ik\left(r_f-r_i\right)}e^{i\Delta T}\left[C(1)H_{A_j}(B(T)) + C(2)_1F_1\left(-\frac{A_j}{2};\frac{1}{2};B^2(T)\right)\right]|j\rangle\langle j||n\rangle \\ e^{-ikr_i}\left[C(1)H_{A_j+1}(B(T)) + C(2)_1F_1\left(-\frac{A_j+1}{2};\frac{1}{2};B^2(T)\right)\right]|j+1\rangle\langle j||n\rangle \end{pmatrix}. \tag{74}$$

As $\langle j|n\rangle = \delta_{j,n}$, we finally obtain an exact and explicit expression for the wave function

$$\Psi(p,g;T) = \sum_{n=0}\omega_n f(p,g,t)$$
$$\times\begin{pmatrix} e^{ik\left(r_f-r_i\right)}e^{i\Delta T}\left[C(1)H_{A_n}(B(T)) + C(2)_1F_1\left(-\frac{A_n}{2};\frac{1}{2};B^2(T)\right)\right]|n\rangle \\ e^{-ikr_i}\left[C(1)H_{A_n+1}(B(T)) + C(2)_1F_1\left(-\frac{A_n+1}{2};\frac{1}{2};B^2(T)\right)\right]|n+1\rangle \end{pmatrix}. \tag{75}$$

This result coincides with that of Ref [20].

5. Conclusions

In this work, we have succeeded in calculating exactly the propagator of the two-level atom interacting with single-mode quantized electromagnetic field and submitted to gravitation using the path integral formalism in the coherent states representation. Thanks to the two angular variables replacing the spin, the propagator has been written, first in the conventional form $\int \mathcal{D}(path)\exp(i/\hbar)S(path)$, then determined exactly. The exactness of the results is displayed in the evaluation of the corresponding wave function. The influence of gravitational waves are reflected in our results.

Author details

Hilal Benkhelil
Laboratoire de physique des rayonnements et de leurs interactions avec la matière (PRIMALAB), Faculté des Sciences de la Matière, Département de Physique, Université Batna 1, Batna, Algeria

*Address all correspondence to: benkhelil.hilal@gmail.com

References

[1] Einstein A, Rosen N. Journal of the Franklin Institute. 1937;**1**:223

[2] Adams C, Sigel M, Mlynek J. Physics Reports. 1994;**240**:143

[3] Kastberg A, Philips W, Rolston S, Spreeuw R, Jessen P. Physical Review Letters. 1995;**74**:1542

[4] Jaynes ET, Cummings F. Proceedings of the IEEE. 1963;**51**:89

[5] Shore BW, Knight PL. Journal of Modern Optics. 1993;**40**:1195

[6] Knight PL, Milonni PW. Physics Reports. 1980;**66**:21

[7] Kien F, Shumovsky AS. International Journal of Modern Physics B. 1991;**5**:2287

[8] Agarwal GS. Journal of the Optical Society of America B: Optical Physics. 1985;**5**:480

[9] Puri RR, Agarwal GS. Physical Review A. 1986;**33**:3610

[10] Barnett SM, Knight PL. Physical Review A. 1986;**33**:2444

[11] Sukumar CV, Buck B. Journal of Physics A. 1984;**17**:885

[12] Singh S. Physical Review A. 1982;**25**: 3206

[13] Si-de D, Gong C-d. Physical Review A. 1994;**50**:779

[14] Sleator T, Wilkens M. Physical Review A. 1993;**48**:3286

[15] Storey P, Collet M, Walls D. Physical Review Letters. 1992;**68**:472

[16] Wang XG, Sun CP. Journal of Modern Optics. 1995;**42**:515

[17] Aouachria M. Canadian Journal of Physics. 2011;**89**:11

[18] Klauder JR, Skagerstam BS. Coherent States. World Scientific: Singapore; 1985

[19] Aouachria M, Chetouani L. Canadian Journal of Physics. 2009;**87**:389

[20] Mohammadi M, Naderi MH, Soltanolkotabi M. Journal of Physics A: Mathematical and Theoretical. 2007; **40**:1377